BIOMETALS IN AUTISM SPECTRUM DISORDERS

BIOMETALS IN AUTISM SPECTRUM DISORDERS

ANDREAS M. GRABRUCKER

Cellular Neurobiology and Neuro-Nanotechnology lab
Department of Biological Sciences, University of Limerick
Castletroy, Limerick, Ireland

ELSEVIER

ACADEMIC PRESS
An imprint of Elsevier

Academic Press is an imprint of Elsevier
125 London Wall, London EC2Y 5AS, United Kingdom
525 B Street, Suite 1650, San Diego, CA 92101, United States
50 Hampshire Street, 5th Floor, Cambridge, MA 02139, United States
The Boulevard, Langford Lane, Kidlington, Oxford OX5 1GB, United Kingdom

Notices

Knowledge and best practice in this field are constantly changing. As new research and experience broaden our understanding, changes in research methods, professional practices, or medical treatment may become necessary.

Practitioners and researchers must always rely on their own experience and knowledge in evaluating and using any information, methods, compounds, or experiments described herein. In using such information or methods they should be mindful of their own safety and the safety of others, including parties for whom they have a professional responsibility.

To the fullest extent of the law, neither the Publisher nor the authors, contributors, or editors, assume any liability for any injury and/or damage to persons or property as a matter of products liability, negligence or otherwise, or from any use or operation of any methods, products, instructions, or ideas contained in the material herein.

Library of Congress Cataloging-in-Publication Data
A catalog record for this book is available from the Library of Congress

British Library Cataloguing-in-Publication Data
A catalogue record for this book is available from the British Library

ISBN: 978-0-12-821132-8

For information on all Academic Press publications visit our website at https://www.elsevier.com/books-and-journals

Publisher: Nikki Levy
Acquisitions Editor: Natalie Farra
Editorial Project Manager: Samantha Allard
Production Project Manager: Kiruthika Govindaraju
Cover Designer: Matthew Limbert

Typeset by TNQ Technologies

Contents

Preface

The physiology and healthy development of our body depend on the interplay of many different factors that interact in and around cells, the functional units of our organ systems. These factors can be grouped depending on their chemical nature, and their abundance and distribution studied in a healthy and diseased background.

All metals together, biometals (the ones needed for biological processes in our body), as well as other metals, form the "metallome" that may also include metalloids (e.g., As, Se, Sb). Thus, the metallome represents the entirety of metal and metalloid species, for example, within a tissue, a cell, or a subcellular compartment (Szpunar, 2004), and therefore metallomes refer to biological systems.

The metallome has also been referred to as ionome (Lahner et al., 2003). The term "ionome" takes into consideration that the ionic state and not the elemental state of the metal is found intra- and extracellular. There, metals may, for example, occur as "free" = aqueous ion, since all metals will be in aqueous solution in our body, or as a protein-bound ion. The potential of metals to bind to ligands is further used to subdivide other "omes" depending on the binding-partner, e.g., the metalloproteome considers the entirety of proteins that bind to metals, and the metallometabolome consists of the entirety of metallometabolites.

The metallome is a rather novel addition to other "omes" that determine the physiology of our body. At the center are, of course, the transcriptome and proteome. The transcriptome represents the entirety of RNA (or mRNA) molecules in one cell or a population of cells, while the proteome is the entire set of proteins that is expressed in a particular compartment at a certain time. While the genome, apart from epigenetic modifications, is rather static and defines the complete set of genes we are born with, the transcriptome and proteome are very flexible and dependent on many factors such as cell type, developmental stage, and even environmental factors.

One of these factors is the metallome and the research field investigating the metallome, and its interaction with other "omes" such as the proteome is referred to as "metallomics" (Haraguchi, 2017). Many proteins need metals to obtain their correct structure or as a factor regulating their enzymatic activity. Indeed, it has been estimated that every third protein requires a metal cofactor (Tainer et al., 1991). Among them are DNA and RNA polymerases that bind zinc (King et al., 2004), and many transcription factors that use a zinc finger motif to interact with the DNA (Cassandri et al., 2017) and control gene expression. Therefore, it is not surprising that the altered availability of metals impacts the transcriptome and, ultimately, the proteome of cells. However, the metallome has further links to other "omes" that have been defined.

Microbiota play an important, and more and more recognized, role in the physiology of our body (see Chapter 12) (Thursby and Juge, 2017). The composition of microbiota our body harbors defines the microbiome. However, as in our own body, cellular processes of microbiota are dependent on the presence of certain metals, and similar metals that are essential for humans are essential for many microbiota as well. Also microbiota such as bacteria are dependent on metals for protein folding, function and cell signaling. In our gut, microbiota are dependent on the supply of essential metals through our diet, and a proportion of metals such as zinc are absorbed by them rather than our cells (Gielda and DiRita, 2012). However, different species have different needs for essential metals or tolerance against toxic metals. Thus, if the metallome of our body changes through lack or overabundance of certain metals, so does very likely the microbiome.

The microbiome, together with the proteome of our cells, has a crucial role in controlling metabolic processes. It, therefore, affects the metabolome, the complete set of small-molecule chemicals found within a biological sample, with its closely related "omes," the glycome (the entire complement of sugars), and the lipidome (the totality of lipids in a compartment) (Fig P.1).

Therefore, the metallome is tightly interconnected with all major determinants ("omes") of healthy cell function. Thus, it should not be surprising that alterations in the metallome, similar to the much more recognized factors causing diseases such as gene mutations affecting the transcriptome and proteome, or the presence of detrimental microbiota, can also affect development and function of our system. Likely, the altered composition of metals will contribute to pathologies by affecting the same pathways identified through disease-associated mutations found in proteins, as undoubtedly many of these proteins will have a metal-binding capacity.

Autism spectrum disorders (ASDs) are a very heterogeneous group of disorders caused by many different factors that have been reported in the past. Many studies investigating large cohorts of patients have shown that

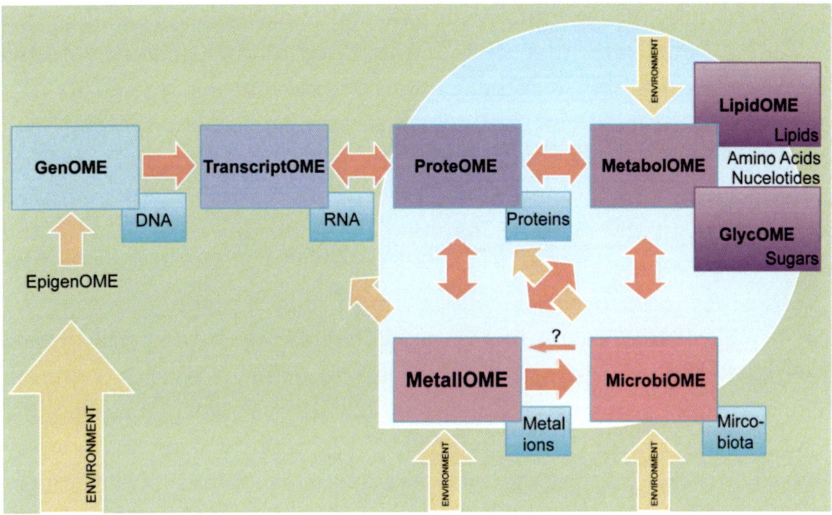

Figure P.1 The genome containing all hereditary information encoded in its DNA is subject to mutations and epigenetic modifications occurring through environmental factors. Mutations in the genome may lead to alterations in the transcriptome and ultimately the proteome. Proteins, as the functional units of cells, orchestrate metabolic processes and thereby are major determinants of the metabolome that, however, is also influenced by environmental factors such as nutrition, toxins, etc., and the presence and activity of microbiota forming the microbiome. The microbiota composition is again under the influence of environmental factors. The metallome is directly or indirectly linked to all other "omes" of our body. The metal composition directly affects gene transcription (transcriptome), and through this, the proteome. The metallome also directly influences protein function of metal-binding proteins and through metals acting as intracellular signaling ions. The abundance of metals also affects the microbiome, and together, alterations in the proteome and microbiome will result in altered metabolic profiles.

the pathogenesis of ASDs has a strong genetic component (Delorme et al., 2013), and several ASD candidate genes were identified by harboring mutations more frequently in individuals with ASDs compared with the control groups. To date, the SFARI (Simons Foundation Autism Research Initiative) gene database lists 888 ASD candidate genes. Fifty-seven are associated with syndromic forms of autism. At the same time, the remaining are classified into six categories according to criteria reflecting the likelihood for them of being causative mutations for ASDs, with 135 genes in category 1 (high confidence), 214 in category 2 (strong candidate), 464 in category 3 (suggestive evidence), and the rest in categories 4–6.

Interestingly, many of the syndromic and high-confidence ASD genes are coding for proteins that are found localized to synapses in the central nervous system (CNS) (Kelleher et al., 2012; Huguet et al., 2016). As a result of this, a model emerged, which shows that synaptic proteins with identified ASD-associated mutations are part of a common signaling pathway that is disrupted in ASDs. Thus, to some extent, ASDs are caused by alterations in the genome, resulting in a different transcriptome and ultimately proteome.

However, despite the identification of this synaptic pathway, the considerable heterogeneity of identified candidate genes that fall outside this model imposes a significant challenge to define a possible common theme. More importantly, ASDs have a strong environmental component as well. Alternative causes (Grabrucker, 2012), where a complex interplay between genetic susceptibility and environmental factors occurs, might account for the majority of ASD cases. For these environmental factors to have an impact, the time window of brain development, starting in utero, is of significant importance. For example, the California Autism Twins Study (CATS) investigating identical and fraternal twins revealed a concordance rate for autism of 31% for male fraternal (heterozygotic) twins and 36% for female fraternal twins (Hallmayer et al., 2011). The concordance rate of fraternal twins was shown to be higher than the observed rate of 3%—14% for siblings of different ages. Thus, despite both sharing, on average, 50% of genes, fraternal twins have a higher concordance rate than siblings. This result indicates that the shared prenatal environment, and therefore the shared exposure to environmental factors in utero, might play a role in the formation of ASDs.

Several environmental risk factors for ASDs have been identified in the past (Grabrucker, 2012). Among them, apart from infections during pregnancy, maternal diabetes, and exposure to certain pesticides and drugs, exposure to toxic metals, prenatal zinc deficiency, and copper overload are the major factors that can be associated with ASDs with high confidence. Unfortunately, a database and scoring system for environmental factors, such as performed for the SFARI gene database, has not been established for environmental causes of ASDs so far. Intriguingly, through the dynamic interaction between the metallome with other "omes," seemingly different causative factors may converge on similar biological targets.

Despite the different causative factors, ASDs can be diagnosed based on the shared features seen in individuals with ASDs. These behavioral alterations, as ultimately all behaviors, have a neurological substrate. Thus, it is reasonable to postulate that at some level, the different causative factors for ASDs will converge and the neurobiological correlate of the observed characteristic behaviors. However, so far, the common motif of the variety of ASD risk factors has not been identified.

It is, therefore, challenging to propose a unified theory of ASDs that encompasses genetic as well as environmental factors and provides a mechanistic explanation for the etiology of ASDs. However, one such model may originate from the interconnectedness of the genome, proteome, transcriptome, metabolome, microbiome, and metallome. As the alterations in the metallome will impact on the transcriptome/proteome, similar biological pathways affected by mutations in specific proteins may be affected by proteins rendered nonfunctional through lack of, or abnormal, metal binding. In addition, a scenario where alterations in the metallome affect the microbiome and by that alter gut–brain signaling through the presence and reaction to bacterial metabolites (altered metabolome), seems plausible.

Therefore, it is timely to investigate closer the role of the metallome in ASDs—not as a separate entity, but as a factor that might contribute together with, and possibly link, already identified genetic and nongenetic factors in the etiology of ASDs. To that end, this book will provide an introduction to the field of metallomics, discuss the evidence amounting so far that links altered metal homeostasis to ASDs, and present insights how metals may affect biological processes linked to ASDs. Especially the essential trace metal zinc and its bioavailability and role in ASDs will be highlighted. Finally, current treatment and prevention strategies that arise from the lessons learned from animal models for ASDs and trace metal abnormalities will be discussed.

Limerick, 01.05.2020

Andreas M. Grabrucker

References

Cassandri, M., Smirnov, A., Novelli, F., Pitolli, C., Agostini, M., Malewicz, M., Melino, G., Raschellà, G., 2017. Zinc-finger proteins in health and disease. Cell Death Discov 3, 17071.

Delorme, R., Ey, E., Toro, R., Leboyer, M., Gillberg, C., Bourgeron, T., 2013. Progress toward treatments for synaptic defects in autism. Nat Med 19 (6), 685–694.

Gielda, L.M., DiRita, V.J., 2012. Zinc competition among the intestinal microbiota. mBio 3 (4) e00171-12.

Grabrucker, A.M., 2012. Environmental factors in autism. Front Psychiatry 3, 118.

Hallmayer, J., Cleveland, S., Torres, A., Phillips, J., Cohen, B., Torigoe, T., Miller, J., Fedele, A., Collins, J., Smith, K., Lotspeich, L., Croen, L.A., Ozonoff, S., Lajonchere, C., Grether, J.K., Risch, N., 2011. Genetic heritability and shared environmental factors among twin pairs with autism. Arch Gen Psychiatry 68 (11), 1095–1102.

Haraguchi, H., 2017. Metallomics: the history over the last decade and a future outlook. Metallomics 9 (8), 1001–1013.

Huguet, G., Benabou, M., Bourgeron, T., 2016. The genetics of autism spectrum disorders. In: Sassone-Corsi, P., Christen, Y. (Eds.), A Time for Metabolism and Hormones. Springer, Cham (CH).

Kelleher, R.J., Geigenmüller, U., Hovhannisyan, H., Trautman, E., Pinard, R., Rathmell, B., Carpenter, R., Margulies, D., 2012. High-throughput sequencing of mGluR signaling pathway genes reveals enrichment of rare variants in autism. PLoS One 7 (4), e35003.

King, R.A., Markov, D., Sen, R., Severinov, K., Weisberg, R.A., 2004. A conserved zinc binding domain in the largest subunit of DNA-dependent RNA polymerase modulates intrinsic transcription termination and antitermination but does not stabilize the elongation complex. J Mol Biol 342 (4), 1143–1154.

Lahner, B., Gong, J., Mahmoudian, M., Smith, E.L., Abid, K.B., Rogers, E.E., Guerinot, M.L., Harper, J.F., Ward, J.M., McIntyre, L., Schroeder, J.I., Salt, D.E., 2003. Genomic scale profiling of nutrient and trace elements in *Arabidopsis thaliana*. Nat Biotechnol 21 (10), 1215–1221.

Szpunar, J., 2004. Metallomics: a new frontier in analytical chemistry. Anal Bioanal Chem 378 (1), 54–56.

Tainer, J.A., Roberts, V.A., Getzoff, E.D., 1991. Metal-binding sites in proteins. Curr Opin Biotechnol 2 (4), 582–591.

Thursby, E., Juge, N., 2017. Introduction to the human gut microbiota. Biochem J 474 (11), 1823–1836.

Acknowledgments

It is a fascinating idea that our body is made, in part, of metals. Unlike Wolverine's adamantium skeleton, our metals are widely distributed in our body, regulating thousands of biological processes that we are just beginning to understand. Luckily, this quest has been supported financially in the recent past. Dr. Grabrucker's research on the contribution of trace metals to autism spectrum disorder (ASD) was funded by the Autism Research Institute (ARI), the Irish Research Council (IRC), the Else Kröner Fresenius Foundation, and the Juniorprofessorenprogramm of the State of Baden-Württemberg, Germany, as well as by start-up funds of the University of Limerick, Ireland. Dr. Grabrucker also received networking support through the EU COST Action TD1304 (Zinc-Net).

Dr. Grabrucker is a faculty member of the Faculty of Science and Engineering, Department of Biological Sciences, at the University of Limerick, and a member of the Bernal Institute of the University of Limerick, and the Health Research Institute (HRI) of the University of Limerick.

CHAPTER 1

Introduction to metallomics: the science of biometals

Contents

Introduction to metallomics

Elements of the periodic table can be classified as nonmetals, metalloids, and metals (Fig. 1.1A). In fact, most of the elements are metals. Elements in this group of metals are further categorized as alkali metals, alkaline earth metals, transition metals, posttransition metals, and the lanthanides and actinides (Fig. 1.1B). At a certain time point in our life, looking at the whole body or different compartments of the body such as organs, cells, or subcellular compartments such as mitochondria, synapses, etc., it is possible to detect a number of these metals that occur in different concentrations. Some of these metals are found because they play a critical role in the physiology of our body, being essential to sustain our life (essential metals, Fig. 1.1C). Others may be found in our bodies as a result of environmental exposure and insufficient mechanisms for export. Together, they build the "metallome," a term coined by Williams, who referred to it as equilibrium concentrations of free metal ions or as a free element content in a cellular compartment, cell, or organism (Williams, 2001).

Metallomics is the study of the metallome, "the entirety of metal and metalloid species present in a cell or tissue type, their identity, quantity and localization" (Mounicou et al., 2009). Dependent on the method used, total metal levels (free ions and protein-bound metals) or only free ions are measured (see Chapter 2).

The metallome is dynamic. Similar to the proteome, where the abundance of specific proteins will negatively or positively affect the levels of

Biometals in Autism Spectrum Disorders
ISBN 978-0-12-821132-8
https://doi.org/10.1016/B978-0-12-821132-8.00001-X

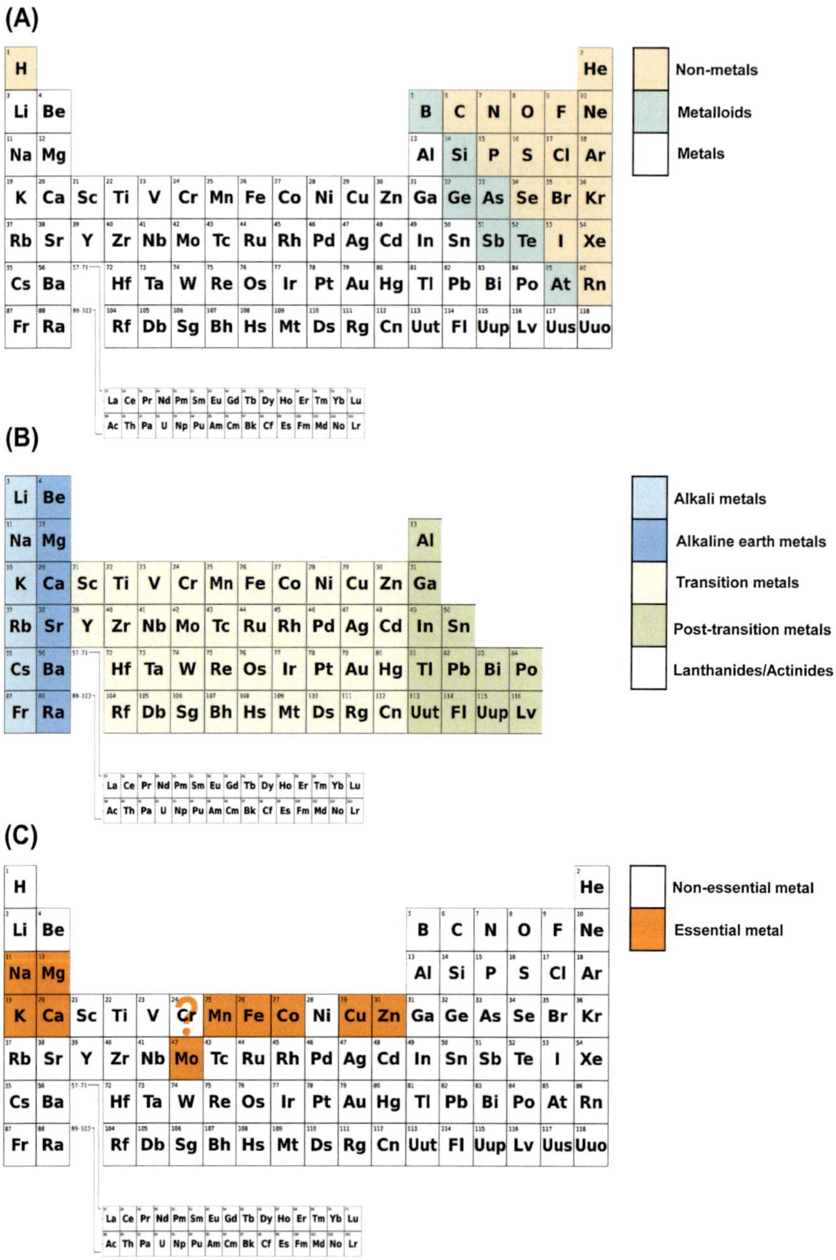

Figure 1.1 (A) The periodic table of elements distinguishes between nonmetals, metalloids, and metals. (B) Metals are further classified as alkali metals, alkaline earth metals, transition metals, posttransition metals, and lanthanides and actinides. (C) Biologically, metals are classified as nonessential and essential metals. The status of Cr is still unclear.

others over time by promoting their degradation, altering rates of mRNA expression and translation, or altering transport, the levels of certain metals will have an impact on the abundance of others. Thus, although metal homeostasis (the metallome) in a cell and tissue is established at some point, the metallome may vary dependent on the agonistic and antagonistic effects of increasing or decreasing a metal and the knock-on effects on others (Fig. 1.2). From this, different stable metal profiles arise, some of which may be characteristic for certain disorders, where the altered metal profiles may be a biomarker but also play a causative or modulatory role (Pfaender and Grabrucker, 2014). ASD, for example, has been frequently associated with a trace metal profile that shows high Cu, Hg, Pb levels, while the levels of Zn, Fe, and Mn are reduced in comparison to individuals without ASD (Pfaender and Grabrucker, 2014).

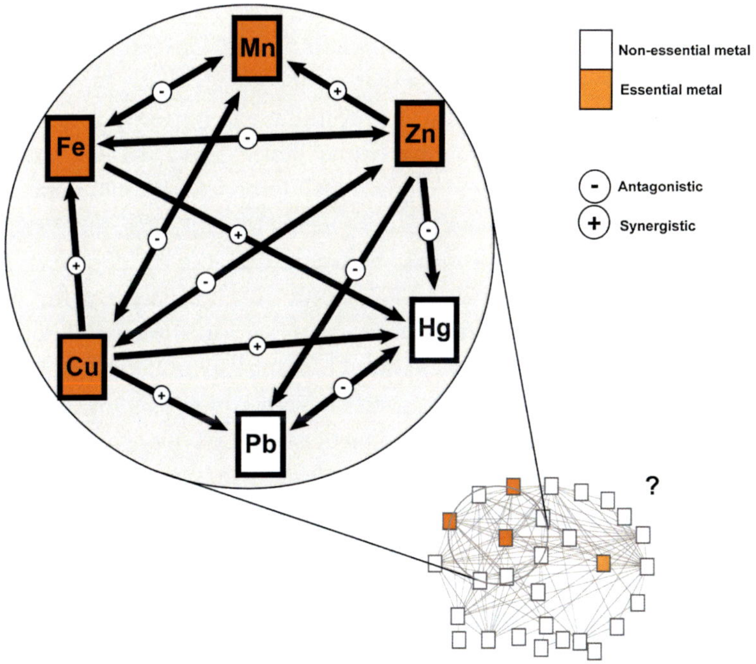

Figure 1.2 Known agonistic and antagonistic effects of the essential trace metals Cu, Fe, Mn, and Zn and the nonessential metals Hg and Pb as an example of the many interactions between metals. With constant metal concentrations, a balance between metals (metal profile) will be established. However, changing metal concentration, that is, an increase in zinc, will not only alter zinc levels, but through interaction, for example, also result in lower Cu, Hg, and Pb levels.

The ambiguity of some proteins mainly causes the agonistic and antagonistic effects of metals in terms of specificity in metal binding. For example, the divalent metal transporter 1 has been shown to mediate the transport of Fe, Zn, Cu, Mn, Cd, Ni, Pb, and Co, all in their 2+ oxidative states (Au et al., 2008; Earley et al., 2008). However, the protein has different affinities for the metals, and the metals compete for binding. Although under normal Fe levels, Fe may be predominantly transported, under low Fe conditions, more Zn or Mn may bind. Similarly, other metal transporters and metal-binding proteins, although showing the highest affinity for a specific metal, may also bind other metals in case the physiological concentration of their preferred metal decreases, or concentrations of other potential metal-ligands increase. Especially, in individuals with ASD, Zn deficiency and Cu overload are frequently reported, and the Zn/Cu ratio was proposed as a biomarker for ASD (Faber et al., 2009).

Zinc and copper, in particular, antagonize each other (Hill and Matrone, 1970; Mills, 1985; Hall et al., 1979) based on their similar physicochemical properties. For example, Cu and Zn form more stable complexes with biological ligands than other metals (e.g., Mg and Mn) (Dudev and Lim, 2014). Because of the tight binding of Cu and Zn to biological metal sites, their intracellular levels have to be carefully controlled. Although there are unique metalloproteins for each metal, such as metallochaperones that deliver Cu to specific binding sites, there are also proteins that mediate the crosstalk between Cu and Zn. For example, proteins of the metallothionein (MT) family that are essential for metal homeostasis and metal detoxification processes are known to bind both Cu and Zn. Thereby, they can buffer the two competitive metal ions.

MTs are small ~7 kDa cellular proteins with a high content of cysteine residues (Drozd et al., 2018). Zinc-binding domains are typically characterized by Zn bound in a tetrahedral geometry to four ligands, most frequently nitrogen from histidine, sulfur from cysteine, and oxygen from aspartate or glutamate (Kochańczyk et al., 2015). MTs fold into two separate domains that can bind Zn and Cu. However, Cu may displace Zn from MTs under some circumstances. In return, the released Zn can then lead to the buffering of a surplus of Cu by activating Cu-binding processes (Krężel and Maret, 2017).

Thus, Zn and Cu levels influence each other, a mechanism exploited as a therapeutic strategy for Wilson's disease, a disorder characterized by abnormal high copper concentrations that can be treated through zinc supplementation (Huster, 2010). Therefore, measuring the metallome is vital for understanding physiological and pathological processes.

Using a variety of approaches, in recent years, information on the metallome of the body and different tissues emerged. However, in comparison to studies using transcriptomics and proteomics, metallomics approaches are significantly underrepresented, and especially correlation analyses between metallome and proteome/microbiome/metabolome are missing. One reason for this is that in comparison to measuring the transcriptome or proteome of a cell, the detection of metals, especially of trace metals and ultratrace metals, is technically much more challenging (see Chapter 2). Nevertheless, recent advances in detection methods lowering the detection threshold for many metals allowed us to gain first insights into the metallome of a healthy body.

The metal composition of our body

Our body, like that of most animated organisms on earth, is composed of only a fraction of all elements found in this universe. Mainly, we are made of oxygen (O), carbon (C), hydrogen (H), nitrogen (N), calcium (Ca), phosphorus (P), potassium (K), sulfur (S), sodium (Na), chlorine (Cl), and magnesium (Mg). These elements alone make up approximately (99.98%) of the human body (Emsley, 1998). However, apart from these common elements, trace elements, despite their relatively low abundance, can be essential for our survival. Among those trace elements, we find many metals such as iron (Fe), zinc (Zn), and copper (Cu). However, with improving instrumentation and methods for the detection of elements, we can identify even more elements in the human body. The role of these ultratrace elements such as gallium (Ga, 70 μg) gold (Au, 20 μg), and vanadium (V, 11 μg in an average human body with 70 kg mass), is currently not well investigated (Chellan and Sadler, 2015). It is possible that some of these ultratrace elements, among them further metals, are completely neutral to the human body. For example, titanium (Ti), of which we find about 20 mg in the average human, is frequently used for medical devices because of its biocompatibility and supposed noninvolvement in biological processes. Others, in turn, may be in balance with metals in our body or affect the human body in ways we do not know yet, while again others such as mercury (Hg) and lead (Pb) have clearly been demonstrated to be toxic even at a minimal concentration (Tchounwou et al., 2012).

What we do know is that certain metals have a clear physiological role in our bodies. Among these so-called biometals are essential trace elements (necessary for survival) and nonessential biometals (metals with a biological role that, however, are not necessary for survival). These biometals are the alkali and alkaline earth metals Na, Mg, K, and Ca (common elements) and the transition metals Fe, Zn, Cu, manganese (Mn), molybdenum (Mo), cobalt (Co), and possibly chromium (Cr). Although Fe (4.2 g/70 kg body weight) and Zn (2.3 g/70 kg) are rather common but mostly called trace elements, Cu (0.072 g/70 kg), Mn (0.012 g/70 kg), Mo (0.005 g/70 kg), and Co (0.003 g/70 kg) are closer to the ultratrace element concentration range (Table 1.1).

Metals in the central nervous system

A significant amount of research points to a central role of pathomechanisms occurring in the brain as a cause of ASD. In particular, the formation, maturation, and maintenance of synapses during brain development and the resulting overall connectivity that is established within and

Table 1.1 Average concentration of transition metals in the human body (Emsley, 1998). Shaded boxes indicate biometals. The status of Cr as biometal is currently not fully clarified.

Transition metals		
Element		mg in 70 kg adult
Fe	Iron	4200
Zn	Zinc	2300
Cu	Copper	72
Ti	Titanium	20
Ni	Nickel	15
Cr	Chromium	14
Mn	Manganese	12
Hg	Mercury	6
Mo	Molybdenum	5
Co	Cobalt	3
Ag	Silver	2
Nb	Niobium	1.5
Zr	Zirconium	1
Y	Yttrium	0.6
Au	Gold	0.2
Sc	Scandium	0.2
Ta	Tantalum	0.2
V	Vanadium	0.11
W	Tungsten	0.02

in between brain regions through synaptic contact formation seem to be central to explain the pathology of ASD (Bourgeron, 2009; Won et al., 2013). Although this does not exclude pathomechanisms outside of the brain, such as immunoregulatory processes or gastrointestinal problems, it is likely that these extracerebral mechanisms will ultimately, for example, through the microbiota–gut–brain axis, influence biological pathways in the brain (Srikantha and Mohajeri, 2019; Cryan et al., 2019). Therefore, it is essential to investigate the metallome of the central nervous system to understand the role of metals in ASD.

However, so far, the metallome of the brain has not been carefully investigated. This can be highlighted through studies aiming at establishing the concentrations of zinc, a key element in ASD, in the brain of humans and likely their most important model organism in biomedical research, mice (Table 1.2).

In general, a wide range of values can be observed in the literature. The variability may be a result of different analytical techniques, human/animal age, brain regions used, or differences in sample preparation. For example, trace metals are rarely measured in fresh human brain tissue. In most studies, the brain tissue has been fixed by formalin treatment or related measures. This procedure usually removes blood within the tissue and, therefore, there is no contribution of trace metals in blood in fixed tissues. In contrast, most studies using animal brain tissue, for example, from rat or mouse, use fresh-frozen tissue. Here, blood is still retained in the tissue, and trace metals in blood will contribute to the final result.

It is therefore complicated to directly compare metallomics studies in a meta-analysis, as even within the same species, different methods and techniques for tissue preparation have been used, and different brain regions measured. Especially, studies assessing young human individuals are scarce due to the unavailability of samples but would be needed to study trace metals in neurodevelopment. Besides, studies using animals, especially animals other than mice and rats, are hardly found in the literature. From the published studies, one can assume a zinc concentration of about 47 mg/kg in brain tissue, and a concentration of around 19 mg/kg in the brain of rodents. However, these differences are likely due to variability in the methods used.

The copper concentrations in the human brain are approx. 6.5 mg/kg (Linder et al., 1998). However, in some brain regions, concentrations can be higher, such as in the substantia nigra, cerebellar cortex, and putamen (Bonilla et al., 1984).

Table 1.2 Comparison of different studies measuring brain zinc levels in human, primate, and rodent brains.

Reference	Species	Brain region	Concentration (mg/kg)	Method
Markesbery et al. (1984)	Human	Whole brain	9.81	INAA
Schrag et al. (2011)	Human	Temporal lobes	28.98 (white matter) 23.66 (gray matter)	AAS
Ramos et al. (2014)	Human	Whole brain Cortex Hippocampus	53.00 61.50 70.00	ICP-MS
Human		Average	41.16 (±18)	
Knauer et al. (2017)	Marmoset	Basal ganglia Hippocampus Thalamus Substantia nigra Superior colliculus	14.70 13.53 14.62 15.41 16.54	ICP-MS
Primate		Average	14.96 (±1.1)	
Pal and Prasad (2016)	Rat	Cortex Striatum Cerebellum	22.30 25.34 32.94	AAS
Yakimaskii et al. (2017)	Rat	Cerebellum	15.00	ICP-MS
Fayed (2010)	Rat	Whole brain	3.13	AAS
Takeda et al. (2000)	Rat	Cortex Hippocampus Cerebellum	11.50 12.50 10.50	AAS
Portbury et al. (2018)	Mouse	Whole brain	12.50	ICP-MS
Harris et al. (2014)	Mouse	Whole brain	15.00	ICP-MS
Yang et al. (2013)	Mouse	Whole brain	16.00	AAS
O'Dell et al. (1989)	Guinea pig	Whole brain	50.00	AAS
Rodent		Average	18.89 (±12.5)	

AAS, atomic absorption spectrometry; ICP-MS, inductively coupled plasma mass spectrometry; INAA, instrumental neutron activation analysis. Concentration in mg/kg brain weight.

Taken together, recent years have seen a great improvement in methodology, which resulted in more profound insights into the metallome of our body and brain, both in health and disease. However, more studies are necessary to understand the dynamics of trace metals and metal composition of our body, especially during hardly accessible time windows such as embryonic and fetal development. In addition, an increase in resolution will be necessary, from the whole organ to the cellular level, to understand the physiological and pathological roles of metals. Finally, future research will have to show how redistribution and alterations of specific metals can be achieved and whether this may be a promising treatment strategy for brain disorders that is mostly untapped so far.

References

Au, C., Benedetto, A., Aschner, M., 2008. Manganese transport in eukaryotes: the role of DMT1. Neurotoxicology 29 (4), 569–576.

Bonilla, E., Salazar, E., Villasmil, J.J., Villalobos, R., Gonzalez, M., Davila, J.O., 1984. Copper distribution in the normal human brain. Neurochem Res 9 (11), 1543–1548.

Bourgeron, T., 2009. A synaptic trek to autism. Curr Opin Neurobiol 19 (2), 231–234.

Chellan, P., Sadler, P.J., 2015. The elements of life and medicines. Philos Trans A Math Phys Eng Sci 373 (2037).

Cryan, J.F., O'Riordan, K.J., Cowan, C.S.M., Sandhu, K.V., Bastiaanssen, T.F.S., Boehme, M., Codagnone, M.G., Cussotto, S., Fulling, C., Golubeva, A.V., Guzzetta, K.E., Jaggar, M., Long-Smith, C.M., Lyte, J.M., Martin, J.A., Molinero-Perez, A., Moloney, G., Morelli, E., Morillas, E., O'Connor, R., Cruz-Pereira, J.S., Peterson, V.L., Rea, K., Ritz, N.L., Sherwin, E., Spichak, S., Teichman, E.M., van de Wouw, M., Ventura-Silva, A.P., Wallace-Fitzsimons, S.E., Hyland, N., Clarke, G., Dinan, T.G., 2019. The microbiota-gut-brain Axis. Physiol Rev 99 (4), 1877–2013.

Drozd, A., Wojewska, D., Peris-Díaz, M.D., Jakimowicz, P., Krężel, A., 2018. Crosstalk of the structural and zinc buffering properties of mammalian metallothionein-2. Metallomics 10 (4), 595–613.

Dudev, T., Lim, C., 2014. Competition among metal ions for protein binding sites: determinants of metal ion selectivity in proteins. Chem Rev 114 (1), 538–556.

Earley, C.J., Ponnuru, P., Wang, X., Patton, S.M., Conner, J.R., Beard, J.L., Taub, D.D., Allen, R.P., 2008. Altered iron metabolism in lymphocytes from subjects with restless legs syndrome. Sleep 31 (6), 847–852.

Emsley, J., 1998. The Elements, third ed. Clarendon Press, Oxford.

Faber, S., Zinn, G.M., Kern, J.C., Kingston, H.M., 2009. The plasma zinc/serum copper ratio as a biomarker in children with autism spectrum disorders. Biomarkers 14 (3), 171–180.

Fayed, A.H., 2010. Brain trace element concentration of rats treated with the plant alkaloid, vincamine. Biol Trace Elem Res 136 (3), 314–319.

Hall, A.C., Young, B.W., Bremner, I., 1979. Intestinal metallothionein and the mutual antagonism between copper and zinc in the rat. J Inorg Biochem 11 (1), 57–66.

Harris, C.J., Voss, K., Murchison, C., Ralle, M., Frahler, K., Carter, R., Rhoads, A., Lind, B., Robinson, E., Quinn, J.F., 2014. Oral zinc reduces amyloid burden in Tg2576 mice. J Alzheimers Dis 41 (1), 179–192.

Hill, C.H., Matrone, G., 1970. Chemical parameters in the study of in vivo and in vitro interactions of transition elements. Fed Proc 29 (4), 1474–1481.

Huster, D., 2010. Wilson disease. Best Pract Res Clin Gastroenterol 24 (5), 531–539.
Knauer, B., Majka, P., Watkins, K.J., Taylor, A.W.R., Malamanova, D., Paul, B., Yu, H.H., Bush, A.I., Hare, D.J., Reser, D.H., 2017. Whole-brain metallomic analysis of the common marmoset (*Callithrix jacchus*). Metallomics 9 (4), 411–423.
Kochańczyk, T., Drozd, A., Krężel, A., 2015. Relationship between the architecture of zinc coordination and zinc binding affinity in proteins—insights into zinc regulation. Metallomics 7 (2), 244–257.
Krężel, A., Maret, W., 2017. The functions of metamorphic metallothioneins in zinc and copper metabolism. Int J Mol Sci 18 (6).
Linder, M.C., Wooten, L., Cerveza, P., Cotton, S., Shulze, R., Lomeli, N., 1998. Copper transport. Am J Clin Nutr 67 (5 Suppl. l), 965–971.
Markesbery, W.R., Ehmann, W.D., Alauddin, M., Hossain, T.I., 1984. Brain trace element concentrations in aging. Neurobiol Aging 5 (1), 19–28.
Mills, C.F., 1985. Dietary interactions involving the trace elements. Annu Rev Nutr 5, 173–193.
Mounicou, S., Szpunar, J., Lobinski, R., 2009. Metallomics: the concept and methodology. Chem Soc Rev 38 (4), 1119–1138.
O'Dell, B.L., Becker, J.K., Emery, M.P., Browning, J.D., 1989. Production and reversal of the neuromuscular pathology and related signs of zinc deficiency in Guinea pigs. J Nutr 119 (2), 196–201.
Pal, A., Prasad, R., 2016. Regional distribution of copper, zinc and iron in brain of wistar rat model for non-wilsonian brain copper toxicosis. Indian J Clin Biochem 31 (1), 93–98.
Pfaender, S., Grabrucker, A.M., 2014. Characterization of biometal profiles in neurological disorders. Metallomics 6 (5), 960–977.
Portbury, S.D., Hare, D.J., Bishop, D.P., Finkelstein, D.I., Doble, P.A., Adlard, P.A., 2018. Trehalose elevates brain zinc levels following controlled cortical impact in a mouse model of traumatic brain injury. Metallomics 10 (6), 846–853.
Ramos, P., Santos, A., Pinto, N.R., Mendes, R., Magalhães, T., Almeida, A., 2014. Anatomical region differences and age-related changes in copper, zinc, and manganese levels in the human brain. Biol Trace Elem Res 161 (2), 190–201.
Schrag, M., Crofton, A., Zabel, M., Jiffry, A., Kirsch, D., Dickson, A., Mao, X.W., Vinters, H.V., Domaille, D.W., Chang, C.J., Kirsch, W., 2011. Effect of cerebral amyloid angiopathy on brain iron, copper, and zinc in Alzheimer's disease. J Alzheimers Dis 24 (1), 137–149.
Srikantha, P., Mohajeri, M.H., 2019. The possible role of the microbiota-gut-brain-axis in autism spectrum disorder. Int J Mol Sci 20 (9).
Takeda, A., Takefuta, S., Okada, S., Oku, N., 2000. Relationship between brain zinc and transient learning impairment of adult rats fed zinc-deficient diet. Brain Res 859 (2), 352–357.
Tchounwou, P.B., Yedjou, C.G., Patlolla, A.K., Sutton, D.J., 2012. Heavy metal toxicity and the environment. Exper Suppl 101, 133–164.
Williams, R.J.P., 2001. Chemical selection of elements by cells. Coord Chem Rev 216–217, 583.
Won, H., Mah, W., Kim, E., 2013. Autism spectrum disorder causes, mechanisms, and treatments: focus on neuronal synapses. Front Mol Neurosci 6, 19.
Yakimoskii, A.F., Shantyr II, Vlasenko, M.A., Yakovleva, M.V., 2017. Effects of acyzol on zinc content in rat brain and blood plasma. Bull Exp Biol Med 162 (3), 293–294.
Yang, Y., Jing, X.P., Zhang, S.P., Gu, R.X., Tang, F.X., Wang, X.L., Xiong, Y., Qiu, M., Sun, X.Y., Ke, D., Wang, J.Z., Liu, R., 2013. High dose zinc supplementation induces hippocampal zinc deficiency and memory impairment with inhibition of BDNF signaling. PLoS One 8 (1), e55384.

CHAPTER 2

Measuring biometals

Contents

Measuring metals

In the past, the determination of metal concentrations has been performed using a variety of different methods and different biological samples such as blood (whole blood, plasma), organ tissue (fixed, fresh, fresh frozen), cerebrospinal fluid (CSF), hair, nail, and teeth. The most commonly used methods are atomic absorption spectroscopy (AAS), inductively coupled plasma—mass spectrometry (ICP-MS), the use of fluorophores, and food frequency questionnaires.

The assessment of metals is a complicated task. Multiple independent factors affect the metal status, including diet and other lifestyle factors such as stress, immune system dysregulation and gastrointestinal health. While some metals are relatively easy to detect, other metals such as zinc (King, 1990) and metals with very low abundance impose significant challenges.

AAS became widely available in the 1960s for accurately measuring trace amounts of mineral in biological samples. Afterward, in the 1970s, inductively coupled plasma—atomic emission spectroscopy (ICP-AES) brought further enhancement in measuring metals more accurately and rapidly. Lately, ICP-MS has been developed, which can measure metals with extremely low detection limits (Miller and Rutzke, 2010).

AAS is one of the most frequently used tools for the determination of most metals (and metalloids). It is based on the absorption of light by free metallic ions and able to detect metals in either liquid or solid samples through their characteristic wavelengths of electromagnetic radiation. To that end, biosamples are reduced into free atoms, for example, using a flame

Biometals in Autism Spectrum Disorders
ISBN 978-0-12-821132-8
https://doi.org/10.1016/B978-0-12-821132-8.00002-1

technique (flame atomic absorption spectroscopy [FAAS]) or electrothermal (graphite tube) atomizers.

Every atomic metal has its distinct pattern at which it will absorb energy (photons of light), leading to a change in the wavelengths that initially passed through the sample. During excitation, electrons of the element move up one energy level, and as electrons return to their original energy state, they emit energy in the form of light. This light has a wavelength that is characteristic of the metal. With increasing concentration of the metal in the sample, absorption will increase proportionally. This allows the qualitative and quantitative analysis of a sample. For quantitative assessments, known concentrations of the metal of interest (standards with known analyte content) are used to generate a calibration curve.

Results from AAS are highly specific, as the resulting atomic absorption lines (emission vs. absorption wavelengths) are rarely overlapping between metals, and several techniques exist to correct for background absorption (Cheremisinoff, 1996; Hill and Fisher, 2017).

ICP-MS has several advantages over AAS, such as the detection of metals in a large linear range, the possibility to detect isotope composition of elements, and, most importantly, very low detection limits. Thus, ICP-MS is a technique that is especially useful to determine low (range: ppb [μg/L]) and ultralow concentrations (ppt [ng/L]) of elements. Besides, ICP-MS has the capability to scan for elements simultaneously. However, the costs for ICP-MS measurements are significantly higher than those of AAS. Besides, matrix effects (interferences caused by sample matrix components) require more calibration strategies compared with AAS.

In brief, ICP-MS uses an ion source, the ICP (typically argon gas, heated with an electromagnetic coil to form plasma). This energy source atomizes a sample (solid or liquid) and transforms the contained elements into ions. Then, the ions created in the plasma are, with the aid of various electrostatic focusing techniques, transmitted through a mass analyzer, counted, and their mass detected. Thus, the ions are separated based on their mass-to-charge ratio. The detector receives an ion signal that is proportional to the concentration of the element, and therefore, as in AAS, a quantitative and qualitative assessment is possible. However, as in AAS, for quantitation, calibration with a known analyte concentration (standards such as single or multi-element references) is necessary (Beauchemin, 2017).

In addition, especially in research, the use of metal-specific fluorophores has allowed the investigation of certain metals with a focus on the tissue and subcellular localization, and dynamics, rather than assessing absolute quantities.

The major disadvantage of AAS and ICP-MS is that a sample, such as a tissue, is atomized in the process of measurement. Therefore, although the metal content and concentration can be precisely determined, the information about the distribution of the metal within the tissue is lost. Also, it is not possible to investigate the redistribution of metals in dynamic processes over time in life cells using AAS or ICP-MS. To overcome this limitation, fluorophores, in combination with high-resolution fluorescent microscopy, have been employed in the past to study the distribution of metals in a biological sample in a spatially resolved manner (Carter et al., 2014).

The basic principle of fluorescence is that a fluorescent molecule can absorb a photon and briefly enter an excited state before returning to its initial state by emitting a photon with lower energy (i.e., longer wavelength). The difference between the wavelengths used to excite the fluorophore and the emission wavelengths (Stokes shift) is used for detection. For example, fluorescent microscopes contain filters that allow light with optimal excitation wavelength to reach the sample but filter these wavelengths for the eye of the observer that is only reached by emission wavelengths. Metal fluorophores are molecules that can bind a specific metal (or several metals but with different affinities), and only metal binding will provide them with the ability to be excited (and thus fluoresce) or lead to a change in the fluorescence intensity. Alternatively, fluorescent molecules may lose their ability to be excited upon binding a metal. Therefore, wherever fluorescence occurs (or is shut off) in tissue after application of a metal fluorophore, a metal ion is present and can be localized (Carter et al., 2014; Calander, 2006) (Fig. 2.1). The fluorescence intensity correlates with the local metal concentration.

Based on these common principles, different classes of fluorescent metal sensors have been developed, for example, molecular probes, genetically encoded probes, and hybrid probes. Molecular probes are comprised of small-molecule fluorophores coupled to a metal-chelating unit. These probes are not synthesized by the cell but need to be delivered, for example, by adding them to the growth medium, application to tissue sections, or microinjected into individual cells (Zhang and Yu, 2008). While some molecular probes are cell permeable, other fluorophores in this class can be made cell permeable through modification with an acetoxymethyl (AM) ester (Tsien, 1981). After entry into a cell, cellular esterases will hydrolyze the AM ester and thereby trap the fluorophore inside the cell.

In contrast, genetically encoded probes are derived from nucleic acid sequences delivered to cells and/or integrated into the cellular DNA and,

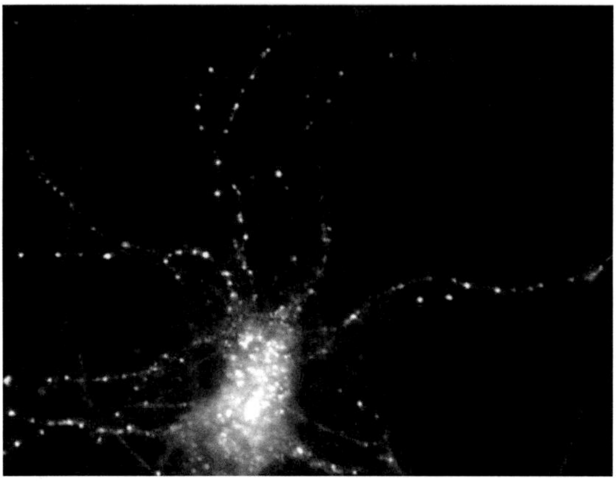

Figure 2.1 Rat hippocampal neurons treated with Zinquin, a fluorophore visualizing zinc. Puncta along the dendrite colocalize with synaptic structures and result from presynaptic free zinc in these "zincergic" neurons and postsynaptic, weakly protein-bound zinc.

therefore, are synthesized by the cellular translation machinery. These intracellular probes are protein based and consist of a fluorescent protein such as GFP fused to a protein domain that has metal-binding capacity. Upon metal binding, a change in the electrochemical environment that leads to a difference in fluorescent signal intensity or a shift in the excitation or emission spectrum is induced (Carter et al., 2014).

Hybrid probes are a combination of molecular and genetically encoded probes. Here, a DNA is introduced in cells that encodes a protein that is, for example, enriched in a specific subcellular compartment. The molecular probe is then covalently attached to the expressed protein and, thus, also enriched at the site where the protein is localized (Keppler et al., 2003, 2004).

Especially the use of calcium fluorophores that allowed calcium imaging, which is a tool to monitor neuronal activity with single-cell resolution, revolutionized the field of neuroscience (Yang and Yuste, 2017). Calcium imaging provides a visual approach to monitoring action potentials in neurons that lead to the spatial redistribution of calcium (Smetter et al., 1999). A wide variety of calcium fluorophores exist that are mainly molecular probes and genetically encoded calcium probes. Commonly used molecular probes are, for example, Fluo-4, Fluo-8, Oregon Green BAPTA-1, Cal-520, Rhod-4, Asante Calcium Red, and X-Rhod-1. Genetically encoded calcium probes are, for example, GCaMP6 variants

(Lock et al., 2015). Because of the applicability of calcium sensors that goes beyond merely visualizing a metal, research on calcium sensor is more advanced than research on fluorophores for any other metal. Calcium sensors can now even detect individual calcium channels in intact cells with a submicrometer spatial resolution (Smith and Parker, 2009). However, in the past, some of the used calcium sensors, such as Calcium Green-1, indeed had a higher affinity for zinc (Stork and Li, 2006).

Besides calcium, zinc ions have recently emerged as a major but so far underinvestigated intracellular and intercellular signal ion. Zinc signaling was studied with several fluorophores in the past. Zinc sensors are available in all three classes of fluorophores. The most commonly used molecular probes are TSQ, Zinquin, FluoZin, ZnAF, Zinpyr1, and Newport Green, but many more are available (for an extensive list, see Carter et al., 2014). Genetically encoded zinc probes are, for example, eCALWY, ZifCY, and ZapCY proteins that are based on zinc finger protein domains or mutated copper-binding domains (van Dongen et al., 2006; Vinkenborg et al., 2009; Dittmer et al., 2009).

Hybrid probes sensing zinc have been based on the so-called SNAP-tag system and the carbonic anhydrase platform. As mentioned earlier, the advantage of hybrid probes is that proteins with specific subcellular locations allow targeting of the small-molecule probes. That way, for example, ZP1, a sensor of the Zinpyr1 family, was conjugated to a protein, which allowed organelle-specific enrichment of the sensor (Juillerat et al., 2003; Tomat et al., 2008).

In addition, an increasing number of fluorophores are available that show a high affinity for other metals. The most common among them are molecular probes such as CTAP-1 and CS1 for copper (Yang et al., 2005; Zeng et al., 2006), RhoNox-1 and IP1 for iron (Au-Yeung et al., 2013), Liang-MN for manganese (Dai et al., 2009), NS1 for nickel (Dodani et al., 2009), CP1 for cobalt (Au-Yeung et al., 2012), LF1 for lead (He et al., 2006), MF1 for mercury (Yoon et al., 2005), and DBITA for cadmium (Liu et al., 2010).

As an alternative to fluorescent-based detection of metals using fluorophores, few histological methods exist that allow the coloring of metals for visualization in light microscopy. For example, already in 1867, the discovery of Perls' Prussian blue allowed staining of nonheme iron (Perls, 1867). Besides, methods such as injection of dithizone visualizing free zinc, or the Timm-Danscher silver enhancement methods that selectively label transition metals (Cu, Zn, Fe, etc.) that are weakly bound are available

(Danscher et al., 1997; Frederickson, 2003). However, all chromogenic dyes that rely on the absorption of light have limited sensitivity when compared with fluorophores.

Measuring metals in biomedical research and the clinics

In clinical practice, measuring the levels of plasma concentrations of metals using AAS such as plasma calcium, iron, and zinc is common (Table 2.1), but these methods have been criticized for their lack of sensitivity, especially for detecting zinc deficiency (King, 1990). Plasma levels of some metals such as zinc can fluctuate greatly throughout the day (by as much as 20%), and therefore, plasma concentrations provide only a snapshot of the zinc status of an individual. This can result in an inaccurate representation of the zinc status (Hambidge et al., 1989). In addition, during blood sample collection, care needs to be taken to avoid hemolysis of blood as red blood cells (RBCs), for example, contain approximately 10 times more zinc than the plasma or serum. Besides, since metals are taken up mainly from dietary sources, blood plasma metal concentration may vary after meals, and therefore, fasting samples are recommended. For example, for children <10 years, the latest threshold concentrations for adequate zinc levels are <9.9 μmol/L (morning, nonfasting) or <8.7 μmol/L (afternoon, nonfasting), and for nonfasting women aged 18–49 years, the values

Table 2.1 Reference values for metals in the blood plasma of fasting adults (male).

	Lower limit	Upper limit	
Ca	~1.05	~1.27	mmol/L
Mg	~0.65	~0.9	mmol/L
Fe	~12.0	~32	μmol/L
Zn	~10.0	~18.5	μmol/L
Cu	~11.0	~24.0	μmol/L
Se	~1.3	~2.0	μmol/L
Mn	~8.0	~20.0	nmol/L
Mo	~5.2	~21.9	nmol/L
Co	~2.4	~8.3	nmol/L
Cr	~2.5	~6.0	nmol/L
Other metals commonly found in plasma			
Al	0	~293	nmol/L
Ni	0	~22.2	nmol/L
Ti	0	~7.9	nmol/L

are <10.1 μmol/L (morning) and <9.0 μmol/L (afternoon). Wessells et al. suggested a cutoff at 7.65 μmol/L to indicate a severe zinc deficiency (Wessels et al., 2014; Wieringa et al., 2015; Tuerk and Fazel, 2009) (Table 2.2). However, as a consequence of fluctuations and sensitivity issues, measuring metal concentrations in hair samples may be more promising but incurs higher costs.

Metal concentrations in hair (or nail) samples represent an average level of the metal present over a more extended period due to the slow deposition in hair and nails and are therefore less affected by rapid fluctuations. Metal in hair or nail samples is commonly determined using AAS (Rahman et al., 2009) or ICP-MS.

Metal levels can also be assessed indirectly by using Food Frequency Questionnaires (FFQ). For example, a study carried out by Trame et al. (2018) developed a biochemically validated questionnaire that can predict the risk of an individual being zinc deficient. In FFQs, the dietary habits of the individuals are analyzed by evaluating the frequency, servings and quantity of different foods consumed over some time (i.e., 6 months). To that end, FFQs usually assess the food type consumed (e.g., red meat, fruits, etc.), the frequency (e.g., daily, one time weekly, etc.), and the portion size (e.g., 1 bar, one slice, one teaspoon, etc.). From that, using known average metal levels in a food type, the amount consumed calculated through translating portion sizes into grams using Food Portion Sizes Databases, and considering the frequency, the metal status of an individual can be assessed. In addition, cross-talk between dietary components and metal absorption can be considered. For example, for the assessment of zinc status, the contribution of zinc from plant sources can be modified (reduced) due to the lower zinc bioavailability in comparison with animal food products (Murphy and Allen, 2003; Sandström et al., 1989). The presence of phytate and fiber results plant-based foods to have less bioavailable zinc (Gibsaon et al., 2018).

Table 2.2 Reference values for trace metals in the blood plasma of fasting babies (0–12 months).

	Lower limit	Upper limit	
Zn	~9.9	~20.0	μmol/L
Cu	~1.4	~20.5	μmol/L
Se	~0.72	~1.21	μmol/L
Mn	~7.3	~50.2	nmol/L

Detecting metal deficiencies

The diagnosis of whether an individual has a metal deficiency or overload is limited by the availability of biosamples, as in most cases, only blood samples will be accessible. Furthermore, diagnosis is hampered by the inability to perform repetitive longitudinal assessments. The latter problem can be resolved by measuring hair samples instead of blood, as hair samples may be obtained easily and accumulate, and thereby average, metal concentrations over several months (depending on the length of the hair). However, the correlation between hair and tissue metal levels is not well established for some metals, and the measurement of hair is often far more expensive as the measurement of blood samples that are routinely assessed using AAS in the clinics.

Although cellular mechanisms exist to keep metal concentrations in our body in a physiological range, the levels of some metals vary over a certain period. While the levels of some metals remain relatively stable, other metals might show high fluctuations. This is highly dependent on the storage and buffering mechanisms. For example, calcium is stored in the extracellular matrix of bones, and the stores are readily accessible in times of low calcium intake to buffer the decrease in blood and tissue calcium levels (Veldurthy et al., 2016). Therefore, calcium levels show only minor fluctuations in blood and other tissues. In contrast, although zinc is also found in bones at relatively high levels, the zinc is not accessible, and thus, bones do not function as zinc stores (King et al., 2000). Therefore, our body has minimal capacity in buffering low zinc intake and in maintaining constant zinc levels. Thus, zinc concentrations in the blood will vary considerably even during a day, depending on the time that past since the last meal and the type of meal. As a result, blood calcium levels are indicative of high or low calcium status in general, but blood zinc levels only represent a snapshot and can only indicate high or low zinc status at the time of blood drawing but are problematic if used to conclude on the zinc status of an individual in general (Wieringa et al., 2015).

Besides, factors such as acute stress, pregnancy, and infections can affect plasma zinc levels (Prasad, 1985). Therefore, the measurement of salivary zinc concentrations has been used in some studies, but the sensitivity of this method has also been questioned (Swanson and King, 1982). Methods such as assessing hair and nail samples for zinc levels could be alternative approaches to measuring the serum or plasma zinc levels. Furthermore, some other parameters used in the assessment of zinc status include Cu:Zn ratio, alkaline phosphatase activity in neutrophils, zinc tolerance test, and the measurement of zinc-dependent enzyme activity (Prasad, 1985). However, unfortunately, there is a lack of available biomarkers that would indicate zinc deficiency in the body (Wieringa et al., 2015).

Moreover, metals occur across a continuous spectrum of concentrations, and therefore, there are no clear boundaries between optimal metal levels, marginal deficiency, severe deficiency, or overload defined in every tissue (Fig. 2.2).

Despite the challenges in measuring certain metals, in the past, several studies measured biometal levels in samples from healthy individuals and compared the results with the measurements performed with samples from individuals with ASD (see Chapters 3 and 6). From these studies, the most frequent finding was an abnormal low zinc concentration observed in samples (hair, nail, blood, teeth) of the group with ASD. Therefore, detecting zinc deficiency is one of the most important challenges in relation to the roles of biometals in ASD.

Measuring zinc

Sensitive, pathologically relevant biomarkers of zinc status are currently lacking. Although several biomarkers have been recently suggested, they are

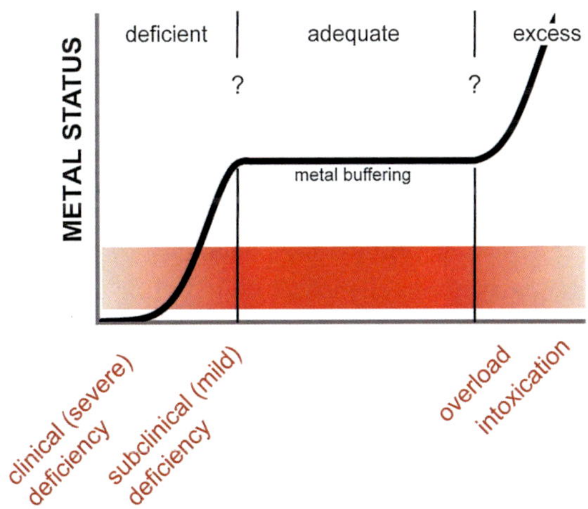

Figure 2.2 Fluctuations in biometal supply can be compensated in a physiological range, although buffering capacities are different for each metal. Outside the physiological range, deficiency or excess occurs. Although limits have been set for each metal, metal concentrations occur along a continuous spectrum, and thus, transitions between adequate metal status, mild and severe deficiency, or overload and intoxication are blurred.

not sufficiently validated to allow their use in clinics as primary assessment methods or diagnostic criterium. Therefore, direct determination of zinc concentrations in plasma, blood cells, nail or hair, or the measurement of urinary zinc excretion needs to be performed. Unfortunately, factors unrelated to zinc status can affect all these measurements, in particular the most frequently used method of determining plasma zinc levels. For example, measuring plasma zinc during pregnancy, the phase of brain development of the embryo that might critically depend on adequate zinc supply is hampered by homeostatic adjustments during pregnancy affecting blood zinc levels (Donangelo and King, 2012).

Therefore, for zinc, several functional indices of zinc status have been suggested. For example, zinc deficiency is associated with diminished taste acuity (hypogeusia). The Zinc Taste Test (ZTT) has been developed to detect zinc deficiency. The ZTT uses 5–10 mL of a 0.1% zinc sulfate solution administered orally to which the reaction of the tested individual is recorded. The participant scores the response on a scale from 1 to 4 (1: no specific taste or other sensation is noticed, even after 10 s; 2: no immediate taste is noted, but after a few seconds a slight taste develops; 3: a definite though not strongly unpleasant taste is noted almost immediately and tends to intensify with time; 4: a strong and unpleasant taste is noted immediately) (Bryce-Smith et al., 1984).

The ZTT was later modified using 2 mL of a 1% solution retained in the mouth for 10 s (Garg et al., 1993) or using 5 mL of 0.1% zinc sulfate solution added to 50 mL of water (for 10 s late at night) (Eaton et al., 2009), and a modified scale from 0 to 4 was introduced. The results obtained using the ZTT have been correlated with serum zinc concentrations during (Garg et al., 1993) and after pregnancy (Nichols et al., 1992), and sweat zinc levels (Eaton et al., 2009). However, a meta-analysis of studies using ZTT concludes that the test lacks sensitivity and specificity to accurately assess zinc status in humans (Gruner and Arthur, 2012).

Later, the ZTT has been further modified, introducing a visual analog scale (VAS) as a measurement of the participant's reaction. The scale is a horizontal line with words that describe the most extreme reaction at both ends. The VAS score is then calculated by measuring the distance from the extreme left to the point along the scale that a tested individual has marked (Zdilla et al., 2015).

Furthermore, light/dark adaptation is affected by zinc deficiency (Nicholson et al., 2014). Although the exact mechanisms are not entirely understood, it was shown that zinc interacts with vitamin A and taurine in the retina and may affect the light—rhodopsin reaction and associated neurotransmission. Besides, plasma levels of retinol-binding protein (RBP) have been reported reduced in zinc deficiency (Nicholson et al., 2014). Therefore, a delayed dark adaptation was proposed to be indicative of zinc status in humans. The test (electroretinography) is performed in a dark room. During the test, a light will blink, and the patient will hit a button to indicate that the light was perceived or the response measured by electrodes. The light will flash at different intensities to test the threshold of the patient's rod function.

Changes in taste acuity and dark adaptation are, however, not exclusively determined by zinc status. Therefore, they cannot be used alone to assess the zinc status of an individual. In addition, the functional tests are particularly problematic for the identification of marginal (subclinical) zinc deficiency.

References

Au-Yeung, H.Y., Chan, J., Chantarojsiri, T., Chang, C.J., 2013. Molecular imaging of labile iron(II) pools in living cells with a turn-on fluorescent probe. J Am Chem Soc 135 (40), 15165—15173.

Au-Yeung, H.Y., New, E.J., Chang, C.J., 2012. A selective reaction-based fluorescent probe for detecting cobalt in living cells. Chem Commun 48 (43), 5268—5270.

Beauchemin, D., 2017. Inductively coupled plasma mass spectromentry methods. In: Lindon, J., Tranter, G.E., Koppenaal, D. (Eds.), Encyclopedia of Spectroscopy and Spectrometry, third ed. Elsevier. 263—246.

Bryce-Smith, D., Simpson, R.I., Southon, S., Johnson, I.T., Gee, J.M., 1984. Anorexia, depression, and zinc deficiency. Lancet 324 (8412), 1162—1163.

Calander, N., 2006. Theory of metal-fluorophore interactions. In: Geddes, C.D., Lakowicz, J.R. (Eds.), Reviews in Fluorescence, Reviews in Fluorescence, vol 2006. Springer, Boston, MA, pp. 71—106.

Carter, K.P., Young, A.M., Palmer, A.E., 2014. Fluorescent sensors for measuring metal ions in living systems. Chem Rev 114 (8), 4564—4601.

Cheremisinoff, P., 1996. Elemental and structural characterization tests. In: Polymer Characterization. Elsevier, pp. 43—83.

Dai, Z., Khosla, N., Canary, J.W., 2009. Visible Colour Displacement Sensing System for Manganese (II). Supramolecular Chemistry. Taylor & Francis.

Danscher, G., Juhl, S., Stoltenberg, M., Krunderup, B., Schrøder, H.D., Andreasen, A., 1997. Autometallographic silver enhancement of zinc sulfide crystals created in cryostat sections from human brain biopsies: a new technique that makes it feasible to demonstrate zinc ions in tissue sections from biopsies and early autopsy material. J Histochem Cytochem 45 (11), 1503—1510.

Dittmer, P.J., Miranda, J.G., Gorski, J.A., Palmer, A.E., 2009. Genetically encoded sensors to elucidate spatial distribution of cellular zinc. J Biol Chem 284 (24), 16289—16297.

Dodani, S.C., He, Q., Chang, C.J., 2009. A turn-on fluorescent sensor for detecting nickel in living cells. J Am Chem Soc 131 (50), 18020—18021.

Donangelo, C.M., King, J.C., 2012. Maternal zinc intakes and homeostatic adjustments during pregnancy and lactation. Nutrients 4 (7), 782—798.

Eaton, K.K., Betteley, I.G., Harris, M., 2009. Diagnosing human zinc deficiency: a comparison between the Bryce-Smith taste test and the sweat mineral analysis. J Nutr Environ Med 14 (2), 83–87.
Frederickson, C., 2003. Imaging zinc: old and new tools. Sci STKE 2003 (182), pe18.
Garg, H.K., Singal, K.C., Arshad, Z., 1993. Zinc taste test in pregnant women and its correlation with serum zinc level. Indian J Physiol Pharmacol 37 (4), 318–322.
Gibson, R.S., Raboy, V., King, J.C., 2018. Implications of phytate in plant-based foods for iron and zinc bioavailability, setting dietary requirements, and formulating programs and policies. Nutr Rev 76 (11), 793–804.
Gruner, T., Arthur, R., 2012. The accuracy of the zinc taste test method. J Alternative Compl Med 18 (6), 541–550.
Hambidge, K.M., Goodall, M.J., Stall, C., Pritts, J., 1989. Post-prandial and daily changes in plasma zinc. J Trace Elem Electrolytes Health Dis 3 (1), 55–57.
He, Q., Miller, E.W., Wong, A.P., Chang, C.J., 2006. A selective fluorescent sensor for detecting lead in living cells. J Am Chem Soc 128 (29), 9316–9317.
Hill, S.J., Fisher, A.S., 2017. Atomic absorption, methods and instrumentation. In: Lindon, J.C., Tranter, G.E., Koppenaal (Eds.), Encyclopedia of Spectroscopy and Spectrometry, third ed. Elsevier, pp. 37–43.
Juillerat, A., Gronemeyer, T., Keppler, A., Gendreizig, S., Pick, H., Vogel, H., Johnsson, K., 2003. Directed evolution of O6-alkylguanine-DNA alkyltransferase for efficient labeling of fusion proteins with small molecules in vivo. Chem Biol 10 (4), 313–317.
Keppler, A., Gendreizig, S., Gronemeyer, T., Pick, H., Vogel, H., Johnsson, K., 2003. A general method for the covalent labeling of fusion proteins with small molecules in vivo. Nat Biotechnol 21 (1), 86–89.
Keppler, A., Pick, H., Arrivoli, C., Vogel, H., Johnsson, K., 2004. Labeling of fusion proteins with synthetic fluorophores in live cells. Proc Natl Acad Sci USA 101 (27), 9955–9959.
King, J.C., 1990. Assessment of zinc status. J Nutr 120 (Suppl. 11), 1474–1479.
King, J.C., Shames, D.M., Woodhouse, L.R., 2000. Zinc homeostasis in humans. J Nutr 130 (5S Suppl. l), 1360–1366.
Liu, Z., Zhang, C., He, W., Yang, Z., Gao, X., Guo, Z., 2010. A highly sensitive ratiometric fluorescent probe for Cd2+ detection in aqueous solution and living cells. Chem Commun 46 (33), 6138–6140.
Lock, J.T., Parker, I., Smith, I.F., 2015. A comparison of fluorescent Ca^{2+} indicators for imaging local Ca^{2+} signals in cultured cells. Cell Calcium 58 (6), 638–648.
Miller, D.D., Rutzke, M.A., 2010. Atomic absorption spectroscopy, atomic emission spectroscopy, and inductively coupled plasma-mass spectrometry. In: Nielsen, S. (Ed.), Food Analysis. Springer, pp. 421–442.
Murphy, S.P., Allen, L.H., 2003. Nutritional importance of animal source foods. J Nutr 133 (11 Suppl. 2), 3932S–3935S.
Nichols, J., Morgan, J., Taylor, A., 1992. Postnatal depression and zinc status—a preliminary study. J Nutr Environ Med 3 (1), 35–42.
Nicholson, B.P., Prasad, A.S., Chew, E.Y., 2014. Zinc deficiency and the eye. In: Preedy, V.R. (Ed.), Handbook of Nutrition, Diet and the Eye. Academic Press, pp. 371–376.
Perls, M., 1867. Nachweis von Eisenoxyd in gewissen pigmenten. Arch für Pathol Anat Physiol für Klin Med 39, 42–48.
Prasad, A.S., 1985. Laboratory diagnosis of zinc deficiency. J Am Coll Nutr 4 (6), 591–598.
Rahman, A., Azad, M.A., Hossain, I., Qusar, M.M., Bari, W., Begum, F., Huq, S.M., Hasnat, A., 2009. Zinc, manganese, calcium, copper, and cadmium level in scalp hair samples of schizophrenic patients. Biol Trace Elem Res 127 (2), 102–108.

Sandström, B., Almgren, A., Kivistö, B., Cederblad, A., 1989. Effect of protein level and protein source on zinc absorption in humans. J Nutr 119 (1), 48–53.
Smetters, D., Majewska, A., Yuste, R., 1999. Detecting action potentials in neuronal populations with calcium imaging. Methods 18 (2), 215–221.
Smith, I.F., Parker, I., 2009. Imaging the quantal substructure of single IP3R channel activity during Ca^{2+} puffs in intact mammalian cells. Proc Natl Acad Sci USA 106 (15), 6404–6409.
Stork, C.J., Li, Y.V., 2006. Intracellular zinc elevation measured with a "calcium-specific" indicator during ischemia and reperfusion in rat hippocampus: a question on calcium overload. J Neurosci 26 (41), 10430–10437.
Swanson, C.A., King, J.C., 1982. Zinc utilization in pregnant and nonpregnant women fed controlled diets providing the zinc RDA. J Nutr 112 (4), 697–707.
Tomat, E., Nolan, E.M., Jaworski, J., Lippard, S.J., 2008. Organelle-specific zinc detection using zinpyr-labeled fusion proteins in live cells. J Am Chem Soc 130 (47), 15776–15777.
Trame, S., Wessels, I., Haase, H., Rink, L., 2018. A short 18 items food frequency questionnaire biochemically validated to estimate zinc status in humans. J Trace Elem Med Biol 49, 285–295.
Tsien, R.Y., 1981. A non-disruptive technique for loading calcium buffers and indicators into cells. Nature 290 (5806), 527–528.
Tuerk, M.J., Fazel, N., 2009. Zinc deficiency. Curr Opin Gastroenterol 25 (2), 136–143.
van Dongen, E.M., Dekkers, L.M., Spijker, K., Meijer, E.W., Klomp, L.W., Merkx, M., 2006. Ratiometric fluorescent sensor proteins with subnanomolar affinity for Zn(II) based on copper chaperone domains. J Am Chem Soc 128 (33), 10754–10762.
Veldurthy, V., Wei, R., Oz, L., Dhawan, P., Jeon, Y.H., Christakos, S., 2016. Vitamin D, calcium homeostasis and aging. Bone Res 4, 16041.
Vinkenborg, J.L., Nicolson, T.J., Bellomo, E.A., Koay, M.S., Rutter, G.A., Merkx, M., 2009. Genetically encoded FRET sensors to monitor intracellular Zn^{2+} homeostasis. Nat Methods 6 (10), 737–740.
Wessells, K.R., King, J.C., Brown, K.H., 2014. Development of a plasma zinc concentration cutoff to identify individuals with severe zinc deficiency based on results from adults undergoing experimental severe dietary zinc restriction and individuals with acrodermatitis enteropathica. J Nutr 144 (8), 1204–1210.
Wieringa, F.T., Dijkhuizen, M.A., Fiorentino, M., Laillou, A., Berger, J., 2015. Determination of zinc status in humans: which indicator should we use? Nutrients 7 (5), 3252–3263.
Yang, L., McRae, R., Henary, M.M., Patel, R., Lai, B., Vogt, S., Fahrni, C.J., 2005. Imaging of the intracellular topography of copper with a fluorescent sensor and by synchrotron x-ray fluorescence microscopy. Proc Natl Acad Sci USA 102 (32), 11179–11184.
Yang, W., Yuste, R., 2017. In vivo imaging of neural activity. Nat Methods 14 (4), 349–359.
Yoon, S., Albers, A.E., Wong, A.P., Chang, C.J., 2005. Screening mercury levels in fish with a selective fluorescent chemosensor. J Am Chem Soc 127 (46), 16030–16031.
Zdilla, M.J., Starkey, L.D., Saling, J.R., 2015. A taste-intensity visual analog scale: an improved zinc taste-test protocol. Integr Med 14 (2), 34–38.
Zeng, L., Miller, E.W., Pralle, A., Isacoff, E.Y., Chang, C.J., 2006. A selective turn-on fluorescent sensor for imaging copper in living cells. J Am Chem Soc 128 (1), 10–11.
Zhang, Y., Yu, L.C., 2008. Single-cell microinjection technology in cell biology. Bioessays 30 (6), 606–610.

CHAPTER 3

The history of metals in autism spectrum disorders

Contents

The history of autism spectrum disorders

The history of metals as a risk factor for ASD is closely linked with the history of ASD and the search for possible causes. In 1943, Leo Kanner first described his classic autistic syndrome (Kanner, 1943) that is characterized by a specific pattern of abnormal behavior. Although he rarely speculated about the causes, he described his "early infantile autism" as an "innate" disorder since he observed a tendency toward autistic behaviors that was confined to some families. He, therefore, may have favored the theory of genetic causes, although, in his later writings, Kanner speculated about the influence of maternal behavior ("refrigerator mothers") with regard to the origin of autism.

While Kanner identified autism in children who mostly had learning disabilities, Hans Asperger reported similar behaviors (the triad of impairments of social interaction, communication, and repetitive patterns of activities) in a different group of children who, however, had cognitive

Biometals in Autism Spectrum Disorders
ISBN 978-0-12-821132-8
https://doi.org/10.1016/B978-0-12-821132-8.00003-3

abilities in the average or superior range (Evans, 2013). Also Asperger showed a tendency to assign hereditary factors to the etiology of autism.

While the contribution of both genetic and environmental factors has been discussed from the very beginning and until now, the advances in the field of genetics and especially in DNA sequencing technology have led to a focus of interest on genetic factors. In January 2020, the Simons Foundation Autism Research Initiative (SFARI) autism gene database (SFARI Gene) lists 913 identified ASD-associated genes grouped in different categories (reflecting the strength of the evidence linking the gene to ASD). Category S ([syndromic disorders] mutations that are not only associated with a high risk for ASD but also consistently linked to additional characteristics not required for an ASD diagnosis) lists 41 genes, category 1 (high confidence) lists 144 genes, category 2 (strong candidate) contains 219 genes, and category 3 (suggestive evidence) lists 472 genes. In addition, 17 copy number variations are listed. These numbers highlight the advances that have been made in undertaking genome-wide association studies. While the human genome project needed 15 months to produce the first draft of the human DNA sequence, with costs amounting to approximately $300 million, today, the cost to generate a whole exome sequence is generally below $1,000, and sequencing takes only a few weeks. Therefore, it is possible to compare the genome of a large number of participants with and without ASD diagnosis, to identify even weakly ASD-associated gene variants (Weiss et al., 2009). These numbers also highlight the wealth of possibilities for research on genetic causes for ASD, which may be one factor for a bias toward overestimating genetic causes.

In contrast, only a handful of definite environmental causes for ASD have been identified over decades (Grabrucker, 2012). Thus, these factors are either mostly responsible for the share of ASD cases without clear genetic cause, or the identification of environmental factors lags because tools and methods to perform research on nongenetic factors have not evolved in the same way as tools and methods (sequencing, generation of knock-out mice) for the study of genetic factors.

Therefore, to this day, there is conflicting data about the contribution of genetic and environmental factors in ASD. While some studies estimate the contribution of genetic factors in the development to be as high as 80% (Sandin et al., 2017), other studies report that genetic factors contribute to only 38% of autism cases, with environmental factors explaining the remaining 62% (Hallmayer et al., 2011). Most likely, the truth lies somewhere in between, and researchers today agree that there are genetic as well as environmental causes. However, in most cases, a combination of both will determine the risk for ASD.

As a consequence of the broadening of the concept of the autism spectrum, estimates of prevalence have increased considerably. However, it is still unclear whether the increase is purely based on heightened awareness and altered diagnostic criteria or whether environmental factors contribute to this increase. Due to the short time in which the increase of ASD diagnoses occurred, the contribution of genetic factors can be ruled out.

In line with this, research on environmental factors has gained traction in recent years. Among the nongenetic risk factors, prenatal infection/immune system disorders, maternal diabetes, pre- and perinatal stress, exposure to certain toxins/pesticides, and abnormal metal homeostasis (zinc deficiency, heavy metal pollution) have the most robust scientific basis (Grabrucker, 2012).

For an environmental factor to act as a risk factor for the development of ASD, the developmental time window of exposure seems to be critical. Several, if not all, nongenetic factors may require in utero exposure. This is supported, for example, by the result of the California Autism Twins Study (Hallmayer et al., 2011). The study showed a concordance rate for ASD that was significantly higher for dizygotic twins than for siblings, although both groups share, on average, 50% of their genes. The higher concordance rate in dizygotic twins can be best explained by the exposure to the same nongenetic factors in the shared maternal environment during pregnancy (Hallmayer et al., 2011).

The history of metal abnormalities in autism spectrum disorders—heavy metal pollution

The first pieces of evidence for abnormal metal homeostasis as a factor in ASD came from early epidemiological observations suspecting the role of toxic heavy metals in the etiology of ASD. Heavy metals are generally defined as metals with high atomic weight, density, or atomic numbers. However, different criteria exist, and the classification of metal as "heavy metal" is not straight forward. While mercury, lead, and bismuth may meet most of the criteria, cadmium, tin, silver, thallium, and even essential trace metals such as copper, iron, and zinc are sometimes considered as heavy metals. Therefore, the term heavy metal pollution is problematic as it refers to a rather unspecific group of metals. In the context of human health and ASD, "heavy metal pollution" points to the presence of abnormally high concentrations of the toxic metals lead (Pb), nickel (Ni), cadmium (Cd), arsenic (As), and mercury (Hg).

Several epidemiological studies in the past have reported that prenatal exposure to heavy metals can cause toxicity that may lead to ASD (Ornoy et al., 2015). In particular, the excess of copper, lead, mercury, and cadmium showed significant associations (Tabatadze et al., 2015). Related to this, fish consumption (increasing mercury exposure), living close to gasoline stations, and the use of aluminum cooking pans have been associated with ASD through maternal exposure (Mohamed Fel et al., 2015). Air pollution by industrial facilities containing arsenic, lead, or mercury has also been linked to increased prevalence of ASD (Dickerson et al., 2015).

Since the middle of the 19th century, heavy metal pollution of the environment through industrial use of heavy metals increased. Although the emissions started to decrease in most developed countries, hot spots of pollution are still presented in highly industrialized areas. Due to the nature of emissions, exposure to heavy metals may occur through a wide range of processes including inhalation and absorption through contaminated water and nutritional sources exposed to contaminated soil (Gorni et al., 2014).

In the past, many studies were performed retrospectively by measuring metals in biosamples of children with ASD. However, depending on the age of the participant and the type of sample, the results not only indicate a postnatal contribution of heavy metals to the development of ASD. Indeed, exposure to heavy metals during early fetal development may be the most critical window of development with regard to ASD.

While blood and urine samples provide only an assessment of acute exposure, hair, nail, and teeth provide historical information of metal exposure. Especially enamel of primary teeth contains a record of prenatal and early postnatal exposure. The development of primary teeth starts in utero, with teeth being completely formed between 3 months and 1 year after birth (Abdullah et al., 2012).

Analyzing suited samples, several studies have shown that children with ASD display a higher body content of toxic metals compared to neurotypical controls (Adams et al., 2009; Eleshehtawy et al., 2011; Lakshmi Priya and Geetha, 2011a). The burden of heavy metals seemed to correlate with the severity of ASD (Geier et al., 2009a, 2009b, 2012). In particular, lead, mercury, and cadmium have been investigated in the past, while few studies with inconsistent results were performed on nickel, tin, or aluminum (Rossignol et al., 2014).

Lead

The presence of lead, even in traces, has effects on the central nervous system, such as deficits in attention, memory, cognition, and behavior (Landrigan et al., 1975; Needleman, 1988). Up until recently, leaded gasoline was the predominant source of airborne lead. In recent years, with the ban of leaded gasoline, industrial emissions account for most of air lead. As a consequence of the past gasoline combustion and present industrial emissions, lead levels may reach high concentrations in soil in some areas. In addition, exposure to lead-based paints contributes to the overall lead burden (Lanphear, 2017; Levin et al., 2008; Chandran; Cataldo, 2010).

Studies performed in the 1990s found that children with ASD had consistently higher blood levels of lead (Cohen et al., 1982; Eppright et al., 1996; Filipek et al., 1999), with 44% of children with ASD having levels greater than two standard deviations above the mean (Filipek et al., 1999). In 2010, although the absolute levels measured in children were lower due to the decreasing lead levels in the environment, research studies still found elevated lead levels in the blood (Clark et al., 2010) or hair and nail samples of children with ASD (Saghazadeh and Rezaei, 2017; Lakshmi Priya; Geetha, 2011a; Elsheshtawy et al., 2011; Fido and Al-Saad, 2005).

These results suggest that both immediate and long-term exposures were higher in children with ASD. However, a detailed metallomics study performed in 2013 using hair samples of different age groups of children with ASD showed that especially the youngest group (infants aged 0–3 years) suffered from a high burden of lead (Yasuda et al., 2013), hinting at an event during and shortly after pregnancy.

Despite several studies reporting a significant correlation between lead levels and the occurrence of ASD, some studies failed to find such a correlation (Albizzati et al., 2012; Adams et al., 2007; Abdullah et al., 2012). Nevertheless, a recent meta-analysis, including the results of 48 independent studies, found significantly increased hair concentrations of lead in individuals with ASD compared to those of control subjects (Saghazadeh and Rezaei, 2017).

Mercury

Mercury that may occur as inorganic (elemental) mercury, and organic mercury is an abundant metal in our environment despite procedures and policies to minimize exposure to it, which are in place in many countries. Most inorganic mercury contamination results from industrial exposure (e.g., waste incinerators, mining activities) and amalgam dental fillings (with the use of dental amalgam rapidly declining). However, organic mercury, such as methylmercury (MeHg), is mostly responsible for mercury exposure of humans (Huang et al., 2012). Humans absorb MeHg mainly through the diet, for example, freshwater and marine fish (Višnjevec et al., 2014).

Increased levels of mercury in children with ASD have been reported in several studies, suggesting they had greater past exposure (Blaurock-Busch et al., 2011; Fida and Al-Saad, 2005). Relating the levels of mercury in hair and nail to the severity of ASD, a positive correlation was found in multiple studies (Laksmi Priya and Geetha, 2011a; Geier et al., 2012).

Again, a focus lies on the pre- and early postnatal developmental time window. Measuring mercury levels in primary whole teeth from both children with ASD and neurotypical children, significantly higher concentrations were found in the ASD group (Adams et al., 2007). Measuring only enamel that is formed later (completely formed only by 3–12 months of age), no significant differences were found (Abdullah et al., 2012), which may support a model of very early (in utero) exposure to be critical.

Increased mercury levels were also reported in blood samples from children with ASD and excessive urinary porphyrin excretion, which is a biomarker of mercury toxicity. Again, mercury levels correlated with the degree of severity of ASD (DeSoto and Hiltan, 2010; Geier et al., 2010, 2009a; 2009b; Nataf et al., 2006; Geier; Geier, 2007; Austin and Shandley, 2008).

Although few studies could not detect higher mercury levels in the blood (Hertz-Picciotto et al., 2010), today, in utero exposure to mercury and lead is established causative risk factors of neurodevelopmental disorders, including ASD (Grandjean and Landrigan, 2006; Rossignol et al., 2014).

Mercury and vaccination

Thimerosal is a *mercury*-based (ethylmercury) compound that has been used for decades as a preservative in several biological and drug products, including many vaccines. Up until 2001, more than 20 vaccines licensed in the United States contained thimerosal. However, the concentrations used per vaccine was very low, in the range of 0.003%–0.01% (Ball et al., 2001). 0.01% thimerosal contains approximately 25 μg of mercury (Hurley et al., 2010). Children receiving all recommended vaccines would still be within the allowable range of mercury stated in the guidelines from the Agency for Toxic Substances and Disease Registry, the Food and Drug Administration, and the World Health Organization (Hurley et al., 2010). Despite this, recommended vaccines for children younger than 6 years in the United

States and other countries no longer contain thimerosal or only trace amounts except for some influenza vaccines (Hurley et al., 2010). Therefore, the amount of mercury in vaccines an infant is exposed to today through vaccination is less than 3 μg per vaccine.

The temporal link between childhood vaccination and the first apparent symptoms of ASD sparked a discussion regarding the possible association between ASD and vaccine administration. As a mode of action, mercury in thimerosal had been suggested, which, causing mercury toxicity, was suspected of leading to ASD. Unfortunately, this theory was supported by a study that contained several elements that were incorrect and contrary to data from earlier investigations. Although the study was retracted due to deliberate fraud (Deer, 2011; Eggerton et al., 2010; Opel et al., 2011), and the theory scientifically disproven, false claims about an association of ASD and vaccination remain present to this day. Multiple epidemiologic studies have failed to show a cause—effect relationship between ASD and vaccination, and studies continue to show no association between thimerosal exposure and ASD. The time window of exposure and concentration of mercury is not matching the studies that show an association of heavy metal pollution with ASD (see above). In these studies, higher levels and an in utero exposure seem to be critical for mercury toxicity to act as a risk factor for the development of ASD.

Cadmium

Cadmium contamination occurs due to different human activities such as waste incineration and the use of phosphate fertilizers (Ciesielski et al., 2012). High levels of cadmium in the soil and water supply may result in absorption through meat, fish, vegetables, and fruits (Rafati Rahimzadeh et al., 2017). Besides, cadmium may enter the body through exposure to tobacco smoke (Fatima et al., 2019).

Some studies have reported an association with high cadmium levels and ASD (Al-Ayadhi, 2005; Blaurock-Busch et al., 2011; Al-Farsi et al., 2013), while one study found lower levels of cadmium in hair samples of children with ASD (Wecker et al., 1985).

A more detailed metallomics analysis using hair samples of children with ASD aged 0—15 years showed that 8.5% display high burdens of cadmium. As the same study reported for lead, the highest body burden was again observed in the infants aged 0—3 years, further hinting at the existence of a critical time window during very early neurodevelopment, likely starting in utero, in which metal imbalances may be particularly detrimental (Yasuda et al., 2013; Yasuda and Tsutsui, 2013).

It is worth noting that high mercury, lead, and cadmium exposure may result from the same processes, such as waste incineration. Individuals with high lead levels will likely, for example, also show high cadmium levels. Therefore, even if only one metal may act as a risk factor for ASD, an association of the others with ASD will be detected in epidemiological studies that mostly reveal correlation but not causation. On the other hand, it may be the combination of these toxic metals that confers ASD risk rather than every single metal alone. Coexposure may affect neurodevelopment through common pathways, and therefore, the effects of these metals may act cumulatively. For example, in humans, concomitant exposure to several metals has more severe and additive effects on cognition than effects observed for each metal alone (Kim et al., 2009; Claus Henn et al., 2012; Rodri; guez-Barranco et al., 2013).

The history of metal abnormalities in autism spectrum disorders—dyshomeostasis of essential metals

The essentiality of the trace element iron has been known for centuries and was finally established in 1932 with the proof that iron was needed for hemoglobin synthesis in humans (Yip and Dallman, 1996). The essentiality of zinc as the second most prevalent trace metal found in our body, in contrast, was only established in 1963 (Prasad, 2013) with little previous notion about its importance for human health. Even less abundant trace metals have been recognized as essential trace metals ahead of zinc. For example, the essentiality of copper for humans was established in 1928 when it was shown to be critical for erythropoiesis (Hart et al., 2002). Manganese was proven to be essential in the 1950s (Emsely, 2013), and the essentiality of cobalt was shown in 1948 with the finding that cobalt functions as a component of vitamin B12 (Rickes et al., 1948).

However, it is zinc that today is primarily in the focus of attention with regard to ASD. Although iron deficiency was also reported to be associated with ASD (Lane et al., 2015), and deficiency in manganese, molybdenum, and selenium was reported, it is zinc deficiency that is a recurring motif that has been confirmed today in a meta-analysis including several epidemiological studies (Pfaender and Grabrucker, 2014).

Iron

Several studies were conducted, which investigated iron levels in individuals with ASD. In some studies, lower iron levels were reported in the ASD group using different biosamples such as blood serum and hair, while other studies did not detect significant differences or report an increase in iron (Table 3.1). For example, analyzing hair of 145 children (65 with ASD and 80 controls) in Saudi Arabia, with a mean age of 8.8 ± 4.0 (ASD) and 7.2 ± 6.3 (CTRL) (Al-Ayadhi, 2005), a significantly lower concentration of iron was found in the hair samples of participants. In contrast, analyzing hair of 54 children (27 with ASD and 27 controls) in Oman, with a mean age of 5.3 ± 1.5 (ASD) and 5.5 ± 1.4 (CTRL) (Al-Farsi et al., 2013), significantly higher hair iron levels were found in the ASD group.

However, the studies differ in the age of participants and the country the study was conducted, which may account for the mixed results. Nevertheless, a recent meta-analysis that investigated 25 studies reporting iron levels of individuals with ASD compared to individuals without ASD similarly found that it is inconclusive whether children with ASD have lower iron levels (Tseng et al., 2018).

Zinc

Most studies measuring zinc in individuals with ASD detect hair zinc levels, as for zinc, hair is the more reliable biosample. Several studies have been performed over the last decades (Table 3.2). While some do

Table 3.1 Selected studies measuring iron in hair and blood (serum ferritin, serum iron, plasma iron) of individuals with ASD and healthy controls (CTRL).

Sample	Participants	Result	Reference
Hair iron	ASD 12, CTRL 22	–	Wecker et al. (1985)
Hair iron	ASD 65, CTRL 80	▼	Al-Ayadhi (2005)
Hair iron	ASD 51, CTRL 40	–	Adams et al. (2006)
Hair iron	ASD 25, CTRL 25	–	Blaurock-Busch et al. (2011)
Hair iron	ASD 44, CTRL 61	–	De Palma et al. (2012)
Hair iron	ASD 27, CTRL 27	▲	Al-Farsi et al. (2013)
Hair iron	ASD 33, CTRL 33	–	Skalny et al. (2017a)
Serum ferritin	ASD 53, CTRL 53	▼	Youssef et al. (2013)
Serum ferritin	ASD 254, CTRL 254	–	Bener et al. (2014)
Serum ferritin	ASD 68, CTRL 18	–	Lane et al. (2015)
Serum iron	ASD 48, CTRL 48	–	Skalny et al. (2017b)
Plasma iron	ASD 20, CTRL 18	▲	Pecorelli et al. (2013)

▼, significantly lower iron levels in ASD group; –, no significant difference; ▲, significantly higher iron levels in ASD group.

Table 3.2 Selected studies measuring zinc in hair, blood (plasma, serum), nail, and teeth of individuals with ASD and healthy controls (CTRL).

Sample	Participants	Result	Reference
Hair	12 ASD, 21 CTRL	▼	Wecker et al. (1985)
Hair	12 ASD, 12 CTRL	–	Shearer et al. (1982)
Hair	51 ASD, 40 CTRL	–	Adams et al. (2006)
Hair	32 ASD, 32 CRTL	▼	Elsheshtawy et al. (2011)
Hair	25 ASD, 25 CRTL	▼	Blaurock-Busch et al. (2011)
Hair	1967 ASD, 719 CTRL	▼	Yasuda et al. (2011)
Hair	27 ASD, 27 CRTL	▲	Al-Farsi et al. (2013)
Hair	30 ASD, 30 CRTL	▼	Tabatadze et al. (2015)
Hair and nail	45 ASD, 50 CTRL	–	Lakshmi Priya and Geetha (2011b)
Plasma	79 ASD, 18 CTRL	–	Russo (2011)
Serum	60 ASD, 60 CTRL	▼ Zn/Cu ratio correlates with ASD severity (CARS)	Li et al. (2014)
Teeth	32 ASD, 44 CTRL	▼	Arora et al. (2017)

▼, significantly lower zinc levels in ASD group; –, no significant difference; ▲, significantly higher zinc levels in ASD group; *CARS*, Childhood Autism Rating Scale.

not report significant differences in zinc levels of the group with ASD compared to healthy controls, the majority reports significantly decreased zinc concentrations. As for iron, a variety of factors might explain the variable results, such as the country where the study took place, the age of the participants, the method used to detect zinc, etc. In particular, the age of the participants may be an important factor. The probably most extensive study to date that investigated 1967 hair samples of individuals with ASD found a clear correlation between age and zinc levels, with the youngest participants (under 3 years of age) having the highest incidence rate of zinc deficiency (43.5% in males and 52.5% in females) (Yasuda et al., 2011).

Two recent meta-analyses confirm that there is a correlation between low zinc status and ASD (Sayehmiri et al., 2015; Babaknejad et al., 2016; Saghazadeh et al., 2017).

Copper

Many studies show that autistic children have higher copper levels compared to healthy controls (Bjorklund, 2013). These can be measured not only in hair but also in blood and nail samples (Table 3.3). Copper is an essential trace metal and cofactor of several enzymes (Davis and Mertz, 1987). However, at high concentrations, copper becomes toxic. Similar to the toxic metals discussed above, increased levels of copper contribute to oxidative stress and disturb the balance of other essential metals, in particular zinc, and also molybdenum (Davis and Mertz, 1987). Copper overload may be a consequence of zinc deficiency but may also be caused by enhanced cytokine levels (enhancing the synthesis of ceruloplasmin in the liver, the major copper-binding copper transport protein) due to immune activation/proinflammatory processes. In the latter case, zinc deficiency could be secondary to copper overload.

In the studies listed above that report abnormal zinc or abnormal copper levels, rarely either zinc or copper is altered. In almost all reports, both, a

Table 3.3 Selected studies measuring copper in hair, nail, blood (plasma, serum), and urine of individuals with ASD and healthy controls (CTRL).

Sample	Participants	Result	Reference
Hair	12 ASD, 12 CTRL	–	Shearer et al. (1982)
Hair	51 ASD, 40 CTRL	–	Adams et al. (2006)
Hair	32 ASD, 32 CRTL	▲ levels correlate with ASD severity (CARS)	Elsheshtawy et al. (2011)
Hair	27 ASD, 27 CRTL	▼	Al-Farsi et al. (2013)
Hair	30 ASD, 30 CRTL	▲	Tabatadze et al. (2015)
Hair and nail	45 ASD, 50 CTRL	▲ levels correlate with ASD severity (CARS)	Lakshmi Priya and Geetha (2011b)
Plasma	79 ASD, 18 CTRL	▲	Russo (2011)
Serum	60 ASD, 60 CTRL	▲ Zn/Cu ratio correlates with ASD severity (CARS)	Li et al. (2014)
Urine	25 ASD, 25 CRTL	▲	Blaurock-Busch et al. (2011)

▼, significantly lower copper levels in ASD group; –, no significant difference; ▲, significantly higher copper levels in ASD group. *CARS*, Childhood Autism Rating Scale.

deficiency in zinc and an increase in copper, occur at the same time. To reflect this, the Zn/Cu ratio may be a more suited parameter to investigate in individuals with ASD. Indeed, even in those studies where one trace metal is not significantly altered, once considering the Zn/Cu ratio, a significant difference between the group with ASD and healthy controls emerges (Lakshmi Priya and Geetha, 2011b). Indeed, an abnormal Zn/Cu ratio has been observed so frequently that is has been proposed as a biomarker for ASD (Faber et al., 2009). Intriguingly, in several studies that investigated an association between the severity of trace metal abnormalities and ASD severity, it was found that the Zn/Cu ratio significantly correlates with ASD severity (the lower the Zn/Cu ratio, the higher the Childhood Autism Rating Scale [CARS] score [high scores hint at more severe cases of ASD]) (Elsheshtawy et al., 2011; Lakshmi Priya; Geetha, 2011b; Li et al., 2014).

Other essential metals

Few studies have measured other trace metals such as manganese, cobalt, and molybdenum levels in individuals with ASD, and meta-analyses are still missing. From the studies that assessed other metals, some report a significant deficiency in cobalt, manganese, and lithium (Table 3.4), while others report high manganese levels. Future studies need to validate these findings.

Taken together, from the plethora of studies investigating metals in individuals with ASD by using different biosamples (hair, nail, blood, teeth), it is safe to conclude that metal homeostasis is significantly disturbed in the ASD group, in particular in young individuals with ASD. Although not all studies show the same results, overall, a clear trend emerges where increased levels of mercury, lead, and copper and decreased levels of zinc are prominent features of the ASD group. Likely, these alterations do not occur independently for each

Table 3.4 Selected studies measuring other metals in individuals with ASD and healthy controls (CTRL).

Sample	Metal	Result	Reference
Hair	Co, Mn	▼	Wecker et al. (1985)
Hair	Li	▼	Adams et al. (2006)
Hair	Mn	▼	Tabatadze et al. (2015)
Teeth	Mn	▼	Arora et al. (2017)

▼, significantly lower levels in the ASD group.

metal, as trace metals establish a specific profile through agonistic and antagonistic interactions (Pfaender and Grabrucker, 2014). However, epidemiological studies cannot answer the question, whether the ASD-specific trace metal profile is cause or consequence of the disorder. Therefore, it is vital to understand the function of metals in brain development.

References

Abdullah, M.M., Ly, A.R., Goldberg, W.A., Clarke-Stewart, K.A., Dudgeon, J.V., Mull, C.G., Chan, T.J., Kent, E.E., Mason, A.Z., Ericson, J.E., 2012. Heavy metal in children's tooth enamel: related to autism and disruptive behaviors? J Autism Dev Disord 42 (6), 929–936.

Adams, J.B., Baral, M., Geis, E., Mitchell, J., Ingram, J., Hensley, A., Zappia, I., Newmark, S., Gehn, E., Rubin, R.A., Mitchell, K., Bradstreet, J., El-Dahr, J.M., 2009. The severity of autism is associated with toxic metal body burden and red blood cell glutathione levels. J Toxicol 2009, 532640.

Adams, J.B., Holloway, C.E., George, F., Quig, D., 2006. Analyses of toxic metals and essential minerals in the hair of Arizona children with autism and associated conditions, and their mothers. Biol Trace Elem Res 110 (3), 193–209.

Adams, J.B., Romdalvik, J., Ramanujam, V.M., Legator, M.S., 2007. Mercury, lead, and zinc in baby teeth of children with autism versus controls. J Toxicol Environ Health A 70 (12), 1046–1051.

Al-Ayadhi, L.Y., 2005. Heavy metals and trace elements in hair samples of autistic children in central Saudi Arabia. Neurosciences 10 (3), 213–218.

Al-Farsi, Y.M., Waly, M.I., Al-Sharbati, M.M., Al-Shafaee, M.A., Al-Farsi, O.A., Al-Khaduri, M.M., Gupta, I., Ouhtit, A., Al-Adawi, S., Al-Said, M.F., Deth, R.C., 2013. Levels of heavy metals and essential minerals in hair samples of children with autism in Oman: a case-control study. Biol Trace Elem Res 151 (2), 181–186.

Albizzati, A., Morè, L., Di Candia, D., Saccani, M., Lenti, C., 2012. Normal concentrations of heavy metals in autistic spectrum disorders. Minerva Pediatr 64 (1), 27–31.

Arora, M., Reichenberg, A., Willfors, C., Austin, C., Gennings, C., Berggren, S., Lichtenstein, P., Anckarsäter, H., Tammimies, K., Bölte, S., 2017. Fetal and postnatal metal dysregulation in autism. Nat Commun 8, 15493.

Austin, D.W., Shandley, K., 2008. An investigation of porphyrinuria in Australian children with autism. J Toxicol Environ Health A 71 (20), 1349–1351.

Babaknejad, N., Sayehmiri, F., Sayehmiri, K., Mohamadkhani, A., Bahrami, S., 2016. The relationship between zinc levels and autism: a systematic review and meta-analysis. Iran J Child Neurol 10 (4), 1–9.

Ball, L.K., Ball, R., Pratt, R.D., 2001. An assessment of thimerosal use in childhood vaccines. Pediatrics 107, 1147–1154.

Bener, A., Khattab, A.O., Al-Dabbagh, M.M., 2014. Is high prevalence of Vitamin D deficiency evidence for autism disorder?: in a highly endogamous population. J Pediatr Neurosci 9 (3), 227–233.

Bjorklund, G., 2013. The role of zinc and copper in autism spectrum disorders. Acta Neurobiol Exp 73 (2), 225–236.

Blaurock-Busch, E., Amin, O.R., Rabah, T., 2011. Heavy metals and trace elements in hair and urine of a sample of arab children with autistic spectrum disorder. Maedica 6 (4), 247–257.

Clark, B., Vandermeer, B., Simonetti, A., Buka, I., 2010. Is lead a concern in Canadian autistic children? Paediatr Child Health 15 (1), 17–22.
Claus Henn, B., Schnaas, L., Ettinger, A.S., Schwartz, J., Lamadrid-Figueroa, H., Hernández-Avila, M., Amarasiriwardena, C., Hu, H., Bellinger, D.C., Wright, R.O., Téllez-Rojo, M.M., 2012. Associations of early childhood manganese and lead coexposure with neurodevelopment. Environ Health Perspect 120 (1), 126–131.
Chandran, L., Cataldo, R., 2010. Lead poisoning: basics and new developments. Pediatr Rev 31 (10), 399–405.
Ciesielski, T., Weuve, J., Bellinger, D.C., Schwartz, J., Lanphear, B., Wright, R.O., 2012. Cadmium exposure and neurodevelopmental outcomes in U.S. children. Environ Health Perspect 120 (5), 758–763.
Cohen, D.J., Paul, R., Anderson, G.M., Harcherik, D.F., 1982. Blood lead in autistic children. Lancet 2 (8289), 94–95.
Davis, G.K., Mertz, W., 1987. Copper. In: Mertz, W. (Ed.), Trace Elements in Human and Animal Nutrition, fifth ed., vol. 1. Academic Press, San Diego, CA, pp. 301–364.
De Palma, G., Catalani, S., Franco, A., Brighenti, M., Apostoli, P., 2012. Lack of correlation between metallic elements analyzed in hair by ICP-MS and autism. J Autism Dev Disord 42 (3), 342–353.
Deer, B., 2011. Secrets of the MMR scare. The Lancet's two days to bury bad news. Br Med J 342, c7001.
Desoto, M.C., Hitlan, R.T., 2010. Sorting out the spinning of autism: heavy metals and the question of incidence. Acta Neurobiol Exp 70 (2), 165–176.
Dickerson, A.S., Rahbar, M.H., Han, I., Bakian, A.V., Bilder, D.A., Harrington, R.A., Pettygrove, S., Durkin, M., Kirby, R.S., Wingate, M.S., Tian, L.H., Zahorodny, W.M., Pearson, D.A., Moyé, L.A., Baio, J., 2015. Autism spectrum disorder prevalence and proximity to industrial facilities releasing arsenic, lead or mercury. Sci Total Environ 536, 245–251.
Eggertson, L., 2010. Lancet retracts 12-year-old article linking autism to MMR vaccines. Can Med Assoc J 182 (4), E199–E200.
Elsheshtawy, E., Tobar, S., Sherra, K., Atallah, S., Elkasaby, R., 2011. Study of some biomarkers in hair of children with autism. Middle East Curr Psychiatr 18, 6–10.
Emsley, J., 2013. Manganese the protector. Nat Chem 5 (11), 978.
Eppright, T.D., Sanfacon, J.A., Horwitz, E.A., 1996. Attention deficit hyperactivity disorder, infantile autism, and elevated blood-lead: a possible relationship. Mo Med 93 (3), 136–138.
Evans, B., 2013. How autism became autism: the radical transformation of a central concept of child development in Britain. Hist Hum Sci 26 (3), 3–31.
Faber, S., Zinn, G.M., Kern, J.C., Kingston, H.M., 2009. The plasma zinc/serum copper ratio as a biomarker in children with autism spectrum disorders. Biomarkers 14 (3), 171–180.
Fatima, G., Raza, A.M., Hadi, N., Nigam, N., Mahdi, A.A., 2019. Cadmium in human diseases: it's more than just a Mere metal. Indian J Clin Biochem 34 (4), 371–378.
Fido, A., Al-Saad, S., 2005. Toxic trace elements in the hair of children with autism. Autism 9 (3), 290–298.
Filipek, P.A., Accardo, P.J., Baranek, G.T., Cook, E.H., Dawson, G., Gordon, B., Gravel, J.S., Johnson, C.P., Kallen, R.J., Levy, S.E., Minshew, N.J., Ozonoff, S., Prizant, B.M., Rapin, I., Rogers, S.J., Stone, W.L., Teplin, S., Tuchman, R.F., Volkmar, F.R., 1999. The screening and diagnosis of autistic spectrum disorders. J Autism Dev Disord 29 (6), 439–484.
Geier, D.A., Audhya, T., Kern, J.K., Geier, M.R., 2010. Blood mercury levels in autism spectrum disorder: is there a threshold level? Acta Neurobiol Exp 70 (2), 177–186.

Geier, D.A., Geier, M.R., 2007. A prospective study of mercury toxicity biomarkers in autistic spectrum disorders. J Toxicol Environ Health A 70 (20), 1723–1730.
Geier, D.A., Kern, J.K., Garver, C.R., Adams, J.B., Audhya, T., Nataf, R., Geier, M.R., 2009a. Biomarkers of environmental toxicity and susceptibility in autism. J Neurol Sci 280 (1–2), 101–108.
Geier, D.A., Kern, J.K., Geier, M.R., 2009b. A prospective blinded evaluation of urinary porphyrins verses the clinical severity of autism spectrum disorders. J Toxicol Environ Health A 72 (24), 1585–1591.
Geier, D.A., Kern, J.K., King, P.G., Sykes, L.K., Geier, M.R., 2012. Hair toxic metal concentrations and autism spectrum disorder severity in young children. Int J Environ Res Publ Health 9 (12), 4486–4497.
Gorini, F., Muratori, F., Morales, M.A., 2014. The role of heavy metal pollution in neurobehavioral disorders: a focus on autism. Rev J Autism Dev Disord 1 (4), 354–372.
Grabrucker, A.M., 2012. Environmental factors in autism. Front Psychiatr 3, 118.
Grandjean, P., Landrigan, P.J., 2006. Developmental neurotoxicity of industrial chemicals. Lancet 368, 2167–2178.
Hallmayer, J., Cleveland, S., Torres, A., Phillips, J., Cohen, B., Torigoe, T., Miller, J., Fedele, A., Collins, J., Smith, K., Lotspeich, L., Croen, L.A., Ozonoff, S., Lajonchere, C., Grether, J.K., Risch, N., 2011. Genetic heritability and shared environmental factors among twin pairs with autism. Arch Gen Psychiatr 68 (11), 1095–1102.
Hart, E.B., Steenbock, H., Waddell, J., Elvehjem, C.A., 2002. Iron in nutrition. VII. Copper as a supplement to iron for hemoglobin building in the rat. 1928. J Biol Chem 277 (34), e22.
Hertz-Picciotto, I., Green, P.G., Delwiche, L., Hansen, R., Walker, C., Pessah, I.N., 2010. Blood mercury concentrations in CHARGE study children with and without autism. Environ Health Perspect 118 (1), 161–166.
Huang, S.S., Strathe, A.B., Fadel, J.G., Lin, P., Liu, T.Y., Hung, S.S., 2012. Absorption, distribution, and elimination of graded oral doses of methylmercury in juvenile white sturgeon. Aquat Toxicol 122–123, 163–171.
Hurley, A.M., Tadrous, M., Miller, E.S., 2010. Thimerosal-containing vaccines and autism: a review of recent epidemiologic studies. J Pediatr Pharmacol Therapeut 15 (3), 173–181.
Kanner, L., 1943. Autistic disturbances of affective contact. Nervous Child 2, 217–250.
Kim, Y., Kim, B.N., Hong, Y.C., Shin, M.S., Yoo, H.J., Kim, J.W., Bhang, S.Y., Cho, S.C., 2009. Co-exposure to environmental lead and manganese affects the intelligence of school-aged children. Neurotoxicology 30 (4), 564–571.
Lakshmi Priya, M.D., Geetha, A., 2011a. A biochemical study on the level of proteins and their percentage of nitration in the hair and nail of autistic children. Clin Chim Acta 412 (11–12), 1036–1042.
Lakshmi Priya, M.D., Geetha, A., 2011b. Level of trace elements (copper, zinc, magnesium and selenium) and toxic elements (lead and mercury) in the hair and nail of children with autism. Biol Trace Elem Res 142 (2), 148–158.
Landrigan, P.J., Whitworth, R.H., Baloh, R.W., Staehling, N.W., Barthel, W.F., Rosenblum, B.F., 1975. Neuropsychological dysfunction in children with chronic low-level lead absorption. Lancet 1 (7909), 708–712.
Lane, R., Kessler, R., Buckley, A.W., Rodriguez, A., Farmer, C., Thurm, A., Swedo, S., Felt, B., 2015. Evaluation of periodic limb movements in sleep and iron status in children with autism. Pediatr Neurol 53 (4), 343–349.
Lanphear, B., 2017. Still treating lead poisoning after all these years. Pediatrics 140 (2).

Levin, R., Brown, M.J., Kashtock, M.E., Jacobs, D.E., Whelan, E.A., Rodman, J., Schock, M.R., Padilla, A., Sinks, T., 2008. Lead exposures in U.S. Children, 2008: implications for prevention. Environ Health Perspect 116 (10), 1285–1293.

Li, S.O., Wang, J.L., Bjørklund, G., Zhao, W.N., Yin, C.H., 2014. Serum copper and zinc levels in individuals with autism spectrum disorders. Neuroreport 25 (15), 1216–1220.

Mohamed Fel, B., Zaky, E.A., El-Sayed, A.B., Elhossieny, R.M., Zahra, S.S., Salah Eldin, W., Youssef, W.Y., Khaled, R.A., Youssef, A.M., 2015. Assessment of hair aluminum, lead, and mercury in a sample of autistic Egyptian children: environmental risk factors of heavy metals in autism. Behav Neurol 2015, 545674.

Nataf, R., Skorupka, C., Amet, L., Lam, A., Springbett, A., Lathe, R., 2006. Porphyrinuria in childhood autistic disorder: implications for environmental toxicity. Toxicol Appl Pharmacol 214 (2), 99–108.

Needleman, H.L., 1988. The persistent threat of lead: medical and sociological issues. Curr Probl Pediatr 18 (12), 697–744.

Opel, D.J., Diekema, D.S., Marcuse, E.K., 2011. Assuring research integrity in the wake of Wakefield. Br Med J 342, d2.

Ornoy, A., Weinstein-Fudim, L., Ergaz, Z., 2015. Prenatal factors associated with autism spectrum disorder (ASD). Reprod Toxicol 56, 155–169.

Pecorelli, A., Leoncini, S., De Felice, C., Signorini, C., Cerrone, C., Valacchi, G., Ciccoli, L., Hayek, J., 2013. Non-protein-bound iron and 4-hydroxynonenal protein adducts in classic autism. Brain Dev 35 (2), 146–154.

Pfaender, S., Grabrucker, A.M., 2014. Characterization of biometal profiles in neurological disorders. Metallomics 6 (5), 960–977.

Prasad, A.S., 2013. Discovery of human zinc deficiency: its impact on human health and disease. Adv Nutr 4 (2), 176–190.

Rafati Rahimzadeh, M., Rafati Rahimzadeh, M., Kazemi, S., Moghadamnia, A.A., 2017. Cadmium toxicity and treatment: an update. Casp J Intern Med 8 (3), 135–145.

Rickes, E.L., Brink, N.G., Koniuszy, F.R., Wood, T.R., Folkers, K., 1948. Vitamin B12, A cobalt complex. Science 108 (2797), 134.

Rodríguez-Barranco, M., Lacasaña, M., Aguilar-Garduño, C., Alguacil, J., Gil, F., González-Alzaga, B., Rojas-García, A., 2013. Association of arsenic, cadmium and manganese exposure with neurodevelopment and behavioural disorders in children: a systematic review and meta-analysis. Sci Total Environ 454–455, 562–577.

Rossignol, D.A., Genuis, S.J., Frye, R.E., 2014. Environmental toxicants and autism spectrum disorders: a systematic review. Transl Psychiatry 4, e360.

Russo, A.J., 2011. Increased copper in individuals with autism normalizes post zinc therapy more efficiently in individuals with concurrent GI disease. Nutr Metab Insights 4, 49–54.

Sandin, S., Lichtenstein, P., Kuja-Halkola, R., Hultman, C., Larsson, H., Reichenberg, A., 2017. The heritability of autism spectrum disorder. J Am Med Assoc 318 (12), 1182–1184.

Saghazadeh, A., Rezaei, N., 2017. Systematic review and meta-analysis links autism and toxic metals and highlights the impact of country development status: higher blood and erythrocyte levels for mercury and lead, and higher hair antimony, cadmium, lead, and mercury. Prog Neuro-Psychopharmacol Biol Psychiatry 79 (Pt B), 340–368.

Saghazadeh, A., Ahangari, N., Hendi, K., Saleh, F., Rezaei, N., 2017. Status of essential elements in autism spectrum disorder: systematic review and meta-analysis. Rev Neurosci 28 (7), 783–809.

Sayehmiri, F., Babaknejad, N., Bahrami, S., Sayehmiri, K., Darabi, M., Rezaei-Tavirani, M., 2015. Zn/Cu levels in the field of autism disorders: a systematic review and meta-analysis. Iran J Child Neurol 9 (4), 1–9.

Shearer, T.R., Larson, K., Neuschwander, J., Gedney, B., 1982. Minerals in the hair and nutrient intake of autistic children. J Autism Dev Disord 12 (1), 25–34.

Skalny, A.V., Simashkova, N.V., Klyushnik, T.P., Grabeklis, A.R., Radysh, I.V., Skalnaya, M.G., Tinkov, A.A., 2017a. Analysis of hair trace elements in children with autism spectrum disorders and communication disorders. Biol Trace Elem Res 177 (2), 215–223.

Skalny, A.V., Simashkova, N.V., Skalnaya, A.A., Klyushnik, T.P., Bjørklund, G., Skalnaya, M.G., Tinkov, A.A., 2017b. Assessment of gender and age effects on serum and hair trace element levels in children with autism spectrum disorder. Metab Brain Dis 32 (5), 1675–1684.

Tabatadze, T., Zhorzholiani, L., Kherkheulidze, M., Kandelaki, E., Ivanashvili, T., 2015. Hair heavy metal and essential trace element concentration IN children with autism spectrum disorder. Georgian Med News 248, 77–82.

Tseng, P.T., Cheng, Y.S., Chen, Y.W., Stubbs, B., Whiteley, P., Carvalho, A.F., Li, D.J., Chen, T.Y., Yang, W.C., Tang, C.H., Chu, C.S., Yang, W.C., Liang, H.Y., Wu, C.K., Yen, C.F., Lin, P.Y., 2018. Peripheral iron levels in children with autism spectrum disorders vs controls: a systematic review and meta-analysis. Nutr Res 50, 44–52.

Višnjevec, A.M., Kocman, D., Horvat, M., 2014. Human mercury exposure and effects in Europe. Environ Toxicol Chem 33 (6), 1259–1270.

Wecker, L., Miller, S.B., Cochran, S.R., Dugger, D.L., Johnson, W.D., 1985. Trace element concentrations in hair from autistic children. J Ment Defic Res 29 (Pt 1), 15–22.

Weiss, L.A., Arking, D.E., Daly, M.J., Chakravarti, A., 2009. A genome-wide linkage and association scan reveals novel loci for autism. Nature 461 (7265), 802–808.

Yasuda, H., Kobayashi, M., Yasuda, Y., Tsutsui, T., 2013. Estimation of autistic children by metallomics analysis. Sci Rep 3, 1199.

Yasuda, H., Tsutsui, T., 2013. Assessment of infantile mineral imbalances in autism spectrum disorders (ASDs). Int J Environ Res Publ Health 10 (11), 6027–6043.

Yasuda, H., Yoshida, K., Yasuda, Y., Tsutsui, T., 2011. Infantile zinc deficiency: association with autism spectrum disorders. Sci Rep 1, 129.

Yip, R., Dallman, P.R., 1996. Iron. In: Ziegler, E.E., Filer, L.J. (Eds.), Present Knowledge in Nutrition, seventh ed. ILSI Press, Washington DC, pp. 278–292.

Youssef, J., Singh, K., Huntington, N., Becker, R., Kothare, S.V., 2013. Relationship of serum ferritin levels to sleep fragmentation and periodic limb movements of sleep on polysomnography in autism spectrum disorders. Pediatr Neurol 49 (4), 274–278.

CHAPTER 4

Essential trace metals and their function in brain development

Contents

Trace metals in brain development and function

Essential trace metals are important for various aspects of brain development and function (summarized in the following). While several specific mechanisms have been identified through targeted manipulation of cell culture systems and animal models, many insights from human studies are based on observational research where frequently several micronutrient deficiencies are present at the same time in investigated individuals. This makes the identification of the individual contribution of a particular trace metal to the overall phenotype difficult. Nevertheless, over recent years, a picture emerged that is able to describe the most prominent roles of each trace metal in brain development and physiology.

The role of iron in brain development and function

Iron is the most prevalent trace metal in the brain. However, a major contribution of iron comes from heme-bound iron found in hemoglobin inside red blood cells within the vast capillary network of the brain. Within the different specific cells of the central nervous system (CNS), the neurons, and glial cells, iron may be less abundant than other trace metals such as zinc. Nevertheless, compared with other organs, the iron concentration in

Biometals in Autism Spectrum Disorders
ISBN 978-0-12-821132-8
https://doi.org/10.1016/B978-0-12-821132-8.00004-5

brain tissue is still relatively high. Iron is vital for the function of mitochondria. It plays a key role in the electron transfer reactions that produce ATP (Upadhyay and Agarwal, 2020). Given that the brain is the organ that consumes high levels of energy (ATP), the concentration of mitochondria, and thus iron, is high compared with other organs. This demand has to be met in the developing brain to ensure proper energy supply. Therefore, during brain development, iron levels increase in several brain regions until about 10 years of age (Höck et al., 1975). However, the concentration of iron in the brain has to be tightly controlled as no excretion mechanisms exist for free iron, and free iron (Fe^{3+}) increases the production of free radicals that may cause oxidative stress.

In addition to its role in the mitochondria, iron is involved in the synthesis of fatty acids that are a vital part of the myelin sheet insulating axons of nerve cells (Ortiz et al., 2004). Thus, iron deficiency significantly affects the production of myelin (Beard, 2003). Furthermore, iron is required for stem cell proliferation and neuronal and glial cell differentiation (Morath and Mayer-Pröschel, 2001). Iron is also involved in the production of several neurotransmitters (serotonin, dopamine adrenalin, and noradrenalin) (Pivina et al., 2019). For example, iron is a cofactor for tyrosine hydroxylase, a key enzyme in the process of dopamine synthesis (Li et al., 2011). In addition, prenatal iron deficiency alters insulin-like growth factor (IGF) signaling in the hippocampus of rats (Tran et al., 2012). IGF-1, in particular, is important for neuron progenitor proliferation and promotes neuron survival and neurite outgrowth, and synapse formation, as well as myelin production in oligodendrocytes (Dyer et al., 2016). Through these and other mechanisms (e.g., mitochondrial function), iron is involved in the processes of synaptogenesis and synaptic plasticity (Muñoz and Humeres, 2012). Therefore, iron is critical for healthy brain development and function.

Iron deficiency leads to anemia (DeLoughery, 2017). Both iron deficiency during pregnancy and the cooccurring anemia has been linked to a decrease in fetal growth and birth weight. In addition, in line with the critical role of iron in the brain, prenatal and perinatal iron deficiency has been associated with impairments in cognition and behavior such as increased anxiety, increased risk for depression, decreased attention spans, a slower speed of neural processing, and more abnormal neurological reflexes, as well as abnormal sleep patterns (Peirano et al., 2010; Lukowski et al., 2010; Lozoff et al., 1998, 2000; Wachs et al., 2005; Amin et al., 2010; Armony-Sivan et al., 2004). The cognitive impairments were evident by significantly lower IQ, poorer language ability, and more deficient fine motor skills of children with low iron status at birth when assessed at 5 years of age (Tamura et al., 2002). Besides, the offspring of women that experienced iron deficiency and anemia during pregnancy had a 3.7-fold greater risk of developing schizophrenia as adults (Insel et al., 2008). However, although some of these symptoms also appear in autism spectrum disorder (ASD), no clear association between prenatal iron deficiency as a causative factor for ASD has been established so far.

The role of zinc in brain development and function

Zinc is the second most prevalent trace metals in the human brain. There, it either occurs as free zinc or is bound to proteins. The vast majority (>80%) is protein bound. However, in contrast to all other trace metals, the number of interaction partners is exceptionally high. About 2800 proteins in the human body bind zinc (Andreini et al., 2006). In addition to these, proteins occur, where zinc can be bound in-between two proteins. Many of these are expressed in the brain. Among the proteins are numerous enzymes (zinc binds to over 300 enzymes (McCall et al., 2000)), transcription factors, and receptors. Proteins bind zinc with different affinities and, therefore, zinc-binding can be transient, and zinc can act as signaling ion for intracellular pathways. In particular, metallothioneins (MTs) are a group of proteins that can transiently bind and release zinc (Maret, 2017). Of the MT family, MT-3 is specifically expressed in the CNS (Hidalgo et al., 2001).

The remaining zinc that is not protein bound is mainly found as a free ion in synaptic vesicles, where it contributes to the processes of neurotransmission (Cole et al., 1999; Palmiter et al., 1996). Thus, through its vast network of zinc-dependent proteins and signaling functions, zinc is involved in numerous processes that are critical for healthy brain development and function (Gower-Winter and Levenson, 2012; Grabrucker, 2016) (Fig. 4.1).

Apart from its general roles in DNA replication, transcriptional control, and mRNA translation that are important for every tissue (Cousins, 1998), zinc plays a more particular role in all stages of brain development. In the earliest phases, zinc availability and signaling is an important factor in cell proliferation, neurogenesis, and neuronal differentiation.

For healthy brain development, both cell proliferation and apoptosis must be tightly controlled. Proper DNA replication is necessary for cell proliferation. DNA polymerases are zinc-binding proteins. Together with zinc-

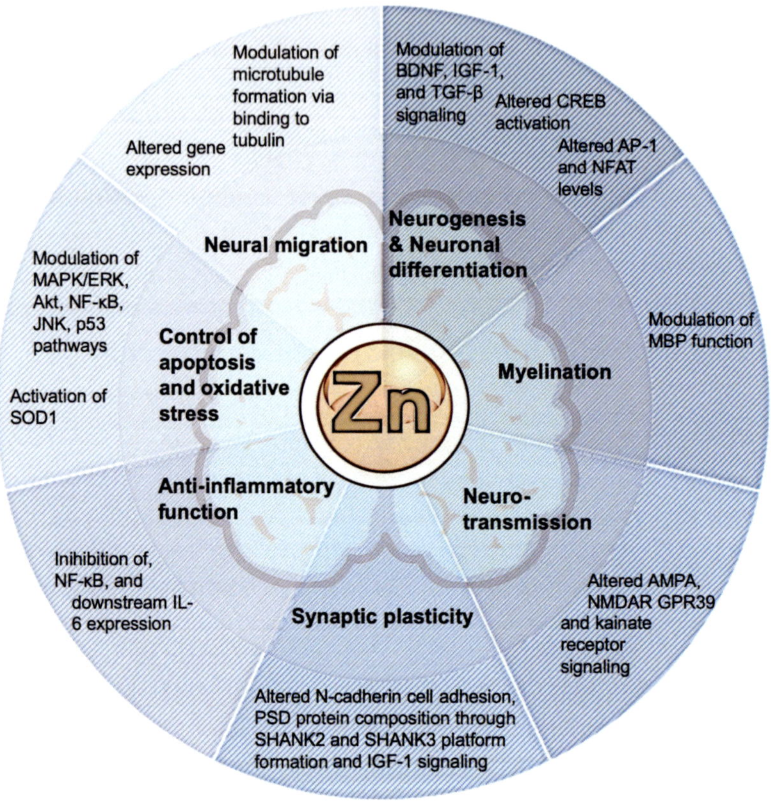

Figure 4.1 ***The roles of zinc in brain development and function.*** Zinc plays a key role in processes of neurogenesis and neuronal differentiation, myelination, neurotransmission and synaptic plasticity, neural migration, as well as the control of the cell cycle and apoptosis. Also, zinc acts antioxidant and antiinflammatory. This is achieved through the modulation of various proteins and critical signaling pathways in the CNS, including, but not limited to, MAPK/ERK, Akt, NF-κB, JNK, p53, NMDAR, AMPAR, BDNF, IGF-1, TGF-β, and SHANK signaling.

dependent transcription factors that regulate the expression of genes necessary to control DNA synthesis, the process of DNA replications is highly zinc dependent. At the same time, proapoptotic processes must be inhibited to ensure the proliferating cells survive. It was shown that zinc has a functional role in this as well. Zinc deficiency not only inhibits cell proliferation through a cell cycle arrest but also induces apoptosis in the early stages of neuronal stem cell formation (Pfaender et al., 2016) and neurodifferentiation (Gao et al., 2009; Adamo et al., 2010). The involved zinc-dependent mechanisms include the modulation of several prosurvival pathways such as the mitogen-activated protein kinase (MAPK)/ERK pathway, the Akt pathway, and the NF-κB (nuclear factor κ-light-chain-enhancer of activated B cells) signaling pathway, as well as proapoptotic pathways such as the c-Jun N-terminal kinase (JNK), and p53 pathways (Pang et al., 2013; Adamo and Oteiza, 2010; Aimo et al., 2010; Seth et al., 2015).

Furthermore, it was shown that several genes involved in cellular proliferation are regulated by zinc in the fetal brain (Aimo et al., 2010), among them the transcription factors AP-1 (activating protein1) and NFAT (nuclear factor of activated T cells). As a result, zinc deficiency during brain development leads to decreased Ki67 and nestin protein levels (Wang et al., 2001; Adamo et al., 2019). Nestin is an intermediate filament protein that is specifically expressed in neural progenitor cells and, therefore, serves as one of the most prominent markers for these cells. The decrease in nestin expression further suggests that zinc deficiency suppresses the development of neural stem cells.

It is known that neurogenesis may also occur in the adult brain. This adult neurogenesis occurs in the subventricular zone of the lateral ventricle wall and the subgranular zone of the dentate gyrus, an input region of the hippocampus. Adult-born dentate granule neurons are functionally integrated into the hippocampal circuitry (Gonçalves et al., 2016). Zinc also plays a role in adult neurogenesis, probably by altering transforming growth factor-beta (TGF-β) signaling (Gower-Winter et al., 2013). A low zinc availability was shown to decrease hippocampal progenitor cell proliferation in mice (Suh et al., 2009).

After neurogenesis, neurons differentiate into different subtypes (e.g., glutamatergic, GABAergic, cholinergic neurons, etc.). Neuronal differentiation is impacted through zinc availability, i.e., controlled by the expression and expression levels of specific zinc transporters regulating zinc (Pfaender et al., 2016). For example, depletion of zinc by genetic deletion of the ZIP12 (SLC39A12), a zinc importer, reduces cAMP response element–binding (CREB) protein activation, which acts as a key molecular switch in a pathway for neuronal differentiation (Chowanadisai et al., 2013). It was shown that human-induced pluripotent stem cell differentiation at low zinc concentrations following a protocol that generates motor neurons results in more cholinergic and less glutamatergic and GABAergic neurons (Pfaender et al., 2016). Similarly, zinc deficiency during brain development resulted in less glutamatergic neurons in the cortex of rats (Adamo et al., 2019).

Newly born neurons often migrate before being incorporated into neuronal networks and circuits. Zinc plays a role in neuronal migration. It was shown that zinc deficiency in rodents impairs neurite outgrowth in the brain (Dvergsten et al., 1984), and cellular zinc signaling was reported to influence the expression of genes implicated in neurite expansion and synaptic growth (Sanford et al., 2019). In addition, zinc also regulates microtubule formation (Hesketh, 1982; Gaskin and Kress, 1977; Oteiza et al., 1990). For example, the migration of cerebellar granular cells is delayed after zinc depletion (Dvergsten et al., 1983).

To ensure proper transduction of electric potentials within neuronal networks, axons of neurons are myelinated by oligodendrocytes in the CNS. Offspring from zinc-deficient rats and monkeys showed a delay in myelin maturation and altered myelin protein profiles (Liu et al., 1992). Zinc is known to bind to and regulate the myelin basic protein (MBP). MBP is an essential protein of the myelin sheath, mediating adhesion of the cytoplasmic membranes of myelin (Bakhti et al., 2014; Tsang et al., 1997).

Another vital part of neurodevelopment and also adult brain function is the process of synapse formation and synaptic plasticity.

Free zinc is found in synaptic neurotransmitter vesicles that also contain glutamate, thus forming a subpopulation of so-called zincergic neurons (Haug, 1967; Holm et al., 1988). The transport of zinc into these vesicles occurs through ZnT3 (SLC30A3) (Cole et al., 1999; Palmiter et al., 1996). Zincergic neurons and their synapses predominantly in the mossy fiber pathway in the hippocampus. However, zincergic neurons can be found in many brain regions, including the cortex, striatum, claustrum, amygdala, and thalamus (Brown and Dyck, 2004; Rockland, 2014; Li and Frederickson, 2013). Almost no zinc-containing neurons are found in the brain stem and cerebellum.

Synaptic zinc acts as a neuromodulator. In fact, it fulfills the criteria of being a neurotransmitter (the substance must be present within the presynaptic neuron, the substance must be released in response to presynaptic depolarization in a calcium-dependent manner, and specific receptors for the substance must be present on the postsynaptic cell) with the exception that zinc is not synthesized in the neuron, which, being an element, is not possible. Thus, after presynaptic depolarization, free zinc is released into the synaptic cleft, which results in a transient zinc signal that may reach concentrations of up to 100–300 μM (Assaf and Chung, 1984; Howell et al., 1984). On the postsynaptic side, zinc binds its specific receptor known as ZnR or GPR39 (Hershfinkel, 2018). In addition, zinc binds to NMDA (*N*-methyl-D-aspartate) and kainate receptors (Mott et al., 2008; Vergnano et al., 2014; Pochwat et al., 2015) in the synaptic cleft and modulates AMPA (α-amino-3-hydroxy-5-methyl-4-isoxazole propionic acid) receptor signaling (Kalappa et al., 2015).

Furthermore, zinc modulates cell adhesion in the synaptic cleft (Wolf et al., 2018) through interacting with N-cadherin (Heiliger et al., 2015). Thereby, zinc may mediate the morphological changes occurring with long-term synaptic plasticity. However, zinc can also enter into the postsynaptic compartment through NMDA and AMPA receptors, and calcium channels (Frederickson et al., 2005) and act as intracellular signaling ion (Fig. 4.2).

Within the postsynaptic neuron, especially the so-called postsynaptic density (PSD), zinc is a key factor regulating the dynamic processes of

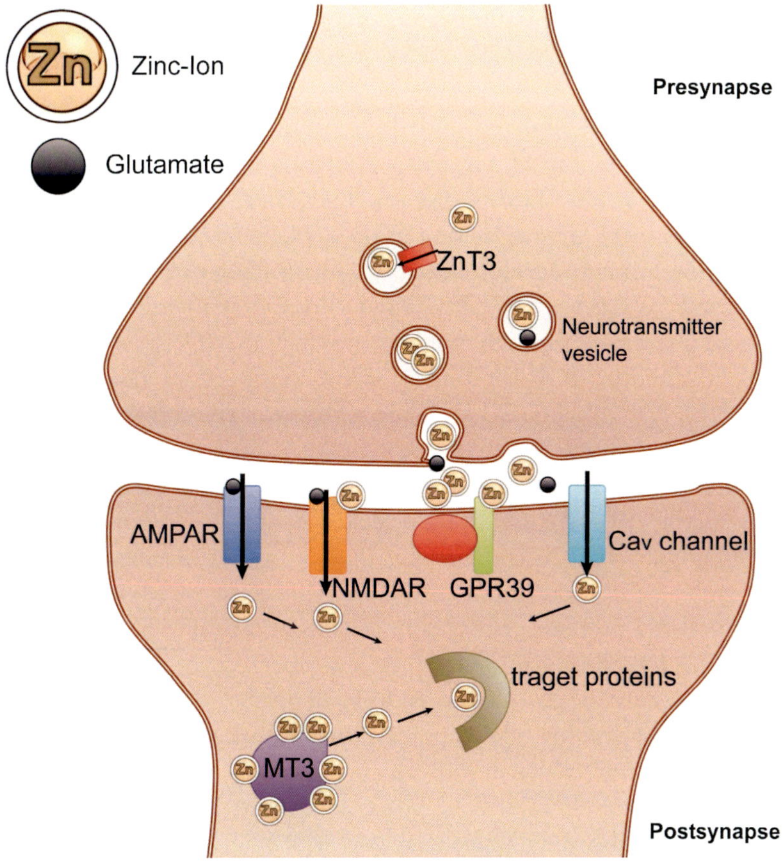

Figure 4.2 Zinc is sequestered into presynaptic glutamate containing neurotransmitter vesicles at zincergic synapses, mediated by ZnT3. It is released into the synaptic cleft upon presynaptic activation. On the postsynaptic side, zinc binds to GPR39 and NMDAR and modulates AMPAR. Zinc can also enter into the postsynaptic compartment through NMDAR, AMPAR, and voltage-gated calcium channels (Ca_V). Inside the postsynapse, zinc can act as signaling ion or bind to target proteins. In addition, zinc can be released in the postsynapse from MT3 in an activity-dependent manner.

protein interactions (Grabrucker, 2014; Jan et al., 2002). It was shown that major scaffold proteins of the PSD of excitatory synapses are zinc binding, and their localization and activity-dependent synaptic functions are driven by zinc (Baron et al., 2006; Grabrucker et al., 2011; Tao-Cheng et al., 2016; Arons et al., 2016; Ha et al., 2018). Intriguingly, these SHANK2 and SHANK3 scaffold proteins are encoded by important ASD candidate genes (Leblond et al., 2014) and at the center of an ASD-associated signaling pathway (Bourgeron, 2009).

Through all these processes mentioned earlier, zinc can influence synapse formation, synaptic transmission, and synaptic plasticity. In line with this, it was shown that the depletion of zinc attenuates the induction of hippocampal long-term potentiation (LTP) (Takeda et al., 2010), a molecular mechanism thought to underlie synaptic plasticity. However, zinc can have a further impact on synapse formation and plasticity through modulation of trophic factors. For example, zinc availability affects the levels of insulin-like growth factor-I (IGF-I) (MacDonald, 2000). IGF-1 has a crucial role in neurodevelopment both prenatally and in the early postnatal period (Wrigley et al., 2017) and contributes to information processing in the brain through synaptic mechanisms (Deak and Sonntag, 2012). In addition, IGF-I promotes adult neurogenesis (Nieto-Estévez et al., 2016). Abnormal IGF-1 signaling was associated with the etiology of ASD (Steinman, 2015, 2019).

Furthermore, a putative zinc—metalloprotease—brain-derived neurotrophic factor (BDNF) axis was suggested to play an essential role in establishing proper brain connectivity (Koh et al., 2014). Abnormal zinc levels in the brain were found to reduce the levels of BDNF, synaptic plasticity-related proteins, and synapse density (Frazzini et al., 2018). As a mechanism, inhibition of zinc-dependent metalloproteases that under physiological conditions, activate BDNF/TrkB signaling have been discussed. The impairment of this process was hypothesized as a factor underlying hyperconnectivity in individuals with ASD (Koh et al., 2014).

Indeed, as a result of several processes that involve zinc binding to proteins or synaptic and intracellular zinc signaling, the establishment of healthy brain connectivity may be impaired. In line with this, it was shown that mice developing under zinc-deficient conditions during pregnancy show altered brain lateralization on molecular, structural, and behavioral levels (Grabrucker et al., 2017).

Besides the critical roles of zinc during brain development and in the adult brain discussed earlier, zinc also has more general roles that may indirectly influence the CNS. For example, zinc can act as an antioxidant. Although zinc is redox inert, zinc binding is important for superoxide dismutase (SOD1) (Hayward et al., 2002). SOD1 is an enzyme that binds copper and zinc to remove superoxide radicals. In addition, zinc can act antiinflammatory. Zinc, for example, through binding to the IκB kinase (IKK) complex, affects NF-κB signaling and thereby may inhibit the expression of inflammatory cytokines such as IL-6 (Liu et al., 2013).

There is a clear link between zinc and ASD. In contrast to other trace metals, the link is established not only through epidemiological studies measuring zinc levels in individuals with ASD but also through mechanistic studies using animal models. Zinc is the only trace metal whose abnormal levels during the brain development of mice result in ASD-like behaviors (Grabrucker et al., 2014, 2016, 2017). In addition, molecular mechanisms have been proposed, elucidating how the reduced availability of zinc, through modulation of SHANK3 signaling, IGF-1, BNDF, and IL-6 signaling, is causing ASD on the molecular level. The proposed mechanisms are in line with processes identified through genetic mutations in ASD, and several key ASD candidate genes encode for zinc-binding protein (for more details, see Chapter 11). In general, while acute zinc deficiency was shown to result in cognitive deficits and depression in mouse models and humans (Hagmeyer et al., 2014), prenatal zinc deficiency is closely linked to ASD.

The role of copper in brain development and function

Copper is the third most abundant trace metal in the human brain, and the brain is the organ with the second highest concentration of copper (after the liver) (Linder et al., 1998). Like zinc, it is an essential cofactor for several enzymes. For example, it is part of cytochrome-c oxidase (Steffens et al., 1987), which is found in mitochondria and is the terminal enzyme of electron transport and oxidative phosphorylation, and superoxide dismutase (SOD) (Robinett et al., 2018), an important regulator of the antioxidative response. Besides, in the brain, copper is found bound to an enzyme that is important for neurotransmitter synthesis, namely dopamine beta-hydroxylase (DBH) (Schmidt et al., 2018). DBH is critical for the conversion of dopamine to norepinephrine. Thus, through DBH, copper plays a role in balancing the levels of the two neurotransmitters.

Apart from acting as cofactor for enzymes, copper also binds to other proteins and thereby plays additional roles, for example, in nerve myelination. Abnormal myelination has been reported as a feature of Menke's disease

(Hara and Taketomi, 1986), a disorder characterized by copper deficiency in liver and brain, and in animal studies using copper-deficient rats (Dipaolo et al., 1974). Although the precise underlying mechanisms are not well understood, altered fatty acid metabolism and impaired mitochondrial energy production may be contributing factors (Gybina et al., 2009).

Demyelination, neurodegeneration, and altered neurotransmitter metabolism, the most likely effects of copper deficiency caused by the altered activity of copper-dependent enzymes and copper-binding proteins, manifest on the behavioral level as motor dysfunction, such as ataxia, and loss of muscular tone. However, copper deficiency has not been linked to ASD-like behavior in human and animal studies.

The role of manganese, cobalt, and molybdenum in brain development and function

Manganese mainly acts as cofactor for enzymes. The most prominent role of these enzymes can be found in processes such as oxidative phosphorylation, fatty acid, cholesterol metabolism, mucopolysaccharide metabolism, and urea cycle (Rehnberg et al., 1982). Some of these processes are also important for brain health and development. However, so far, manganese deficiency is not associated with prominent neuropathological features other than enhanced susceptibility to seizures (Takeda, 2003), although manganese deficiency was found in individuals with ASD. In contrast, manganese overload, among others, causes apathy, headache, speech disturbance, psychosis, and parkinsonian-like syndrome (Van Rij et al., 1979). Thus, while manganese overload is better understood on the molecular level, the physiological role of manganese is less well investigated. What is known is that manganese is mostly bound to proteins in the brain, in particular the enzyme glutamine synthetase (GS), which is predominantly found in astrocytes (Montes et al., 2003). There, GS is one of several enzymes that regulate glutamate/glutamine and GABA homeostasis in the brain. Manganese inhibits GS. Given that astrocytes are distributing glutamine to neurons (Sidoryk-Wegrzynowicz and Aschner, 2013), a process that is under the control of neurons (Hertz et al., 1999), inhibition of GS lowers glutamate availability for neurotransmission, where glutamate acts as the primary excitatory neurotransmitter. Furthermore, manganese is bound to Mn superoxide dismutase enzymes that play a role in the regulation of oxidative stress, and arginase, enzymes that are also expressed in the CNS (Horning et al., 2015).

In addition, it has been speculated that manganese may also be sequestered in synaptic vesicles and released into the synaptic cleft upon synaptic activity (Takeda et al., 2002). However, the physiological role of synaptic manganese is currently not known.

Cobalt is a component of cobalamin (vitamin B12). Cobalamin deficiency has been linked to cardiomyopathy, congestive cardiac failure, pericardial effusion, polycythemia, and thyroid enlargement (Barceloux, 1999). The symptoms emerging from cobalamin deficiency that can be attributed to altered CNS function are memory impairment, attention deficits, dementia, and, less often, optic nerve atrophy (Briani et al., 2013). In the brain, cobalamin acts as a cofactor in the methylmalonyl-CoA mutase reaction, which is a critical step in the process of myelin synthesis. In this reaction, the conversion of methylmalonyl-CoA to succinyl-CoA takes place. Low levels of cobalamin, therefore, lead to the accumulation of methylmalonyl-CoA. It has been reported that this accumulation causes a decrease in normal myelin synthesis and incorporation of abnormal fatty acids into lipids necessary for myelin production (Scalabrino, 2009). Thus, cobalamin deficiency results in abnormal myelination. Cobalamin deficiency also affects monoamine neurotransmitter production as cobalamin (and folate) stimulates tetrahydrobiopterin (BH4) synthesis. BH4 is an essential cofactor in the monoamine neurotransmitter biosynthesis (Hutto, 1997; Miller, 2008). Together, these mechanisms may contribute to the neurological symptoms of cobalamin deficiency. However, many effects on the brain or brain development may also occur secondary, for example, through altered thyroid hormone levels (Eng and Lam, 2020; Admati et al., 2020). Cobalamin deficiency has been linked to ASD (Zhang et al., 2016). It has been speculated that cobalamin-related redox metabolic pathways in neuronal cells are dysregulated, and the methylation capacity decreased (Zhang et al., 2016). However, cobalamin supplementation showed mixed results in children with ASD (Bertoglio et al., 2010; Hendren et al., 2016). Additionally, very high cobalamin levels during pregnancy were associated with an increased risk for ASD (Raghavan et al., 2018).

Molybdenum is part of a complex called molybdenum cofactor that is required for three enzymes in humans: xanthine oxidase (XO), aldehyde oxidase (AO), and sulfite oxidase (SO) (Sardesai, 1993). For the brain, especially SO seems to be necessary, as deficiencies of SO or molybdenum cofactor results in mental retardation and neurologic problems. Most likely, these abnormalities are caused by the accumulation of toxic sulfite and/or inadequate amounts of inorganic sulfate in the brain due to the loss of SO function (Mendel and Bittner, 2006). However, molybdenum deficiency is extremely rare, and most insights come from genetic mutations in the genes encoding for the three enzymes rather than molybdenum deficient individuals. Molybdenum deficiency has also not been linked to ASD-like symptoms.

Conclusions

While epigenetic studies showed altered trace metal profiles associated with ASD, a major question remains, whether the alterations are a cause or consequence, and whether the alterations of all metals together are associated with ASD, or whether the altered level of a single metal is the key risk factor. To answer this question, the individual roles of the trace metals in brain development and function need to be understood on the molecular level, and mechanistic studies using model systems need to be performed. From these studies, currently, zinc seems to be the trace metal that shows the most direct connection to observed pathologies in ASD. Indeed, zinc deficiency during brain development affects signaling pathways that are already known to play a significant role in the etiology of ASD through genetic mutations in key pathway proteins that have been identified in human individuals with ASD. However, the homeostatic mechanisms that regulate the levels of a particular trace metal are influenced by the presence and abundance of other metals. Thus, although zinc deficiency may ultimately trigger processes that increase the risk for ASD, other metals, but also genetic as well as nongenetic factors, might act upstream influencing zinc availability.

References

Adamo, A.M., Liu, X., Mathieu, P., Nuttall, J.R., Supasai, S., Oteiza, P.I., 2019. Early developmental marginal zinc deficiency affects neurogenesis decreasing neuronal number and altering neuronal specification in the adult rat brain. Front Cell Neurosci 13, 62.

Adamo, A.M., Oteiza, P.I., 2010. Zinc deficiency and neurodevelopment: the case of neurons. Biofactors 36 (2), 117−124.

Adamo, A.M., Zago, M.P., Mackenzie, G.G., Aimo, L., Keen, C.L., Keenan, A., Oteiza, P.I., 2010. The role of zinc in the modulation of neuronal proliferation and apoptosis. Neurotox Res 17 (1), 1−14.

Admati, I., Wasserman-Bartov, T., Tovin, A., Rozenblat, R., Blitz, E., Zada, D., Lerer-Goldshtein, T., Appelbaum, L., 2020. Neural alterations and hyperactivity of the hypothalamic-pituitary-thyroid axis in Oatp1c1 deficiency. Thyroid 30 (1), 161−174.

Aimo, L., Mackenzie, G.G., Keenan, A.H., Oteiza, P.I., 2010. Gestational zinc deficiency affects the regulation of transcription factors AP-1, NF-κB and NFAT in fetal brain. J Nutr Biochem 21 (11), 1069−1075.

Amin, S.B., Orlando, M., Eddins, A., MacDonald, M., Monczynski, C., Wang, H., 2010. In utero iron status and auditory neural maturation in premature infants as evaluated by auditory brainstem response. J Pediatr 156 (3), 377–381.
Andreini, C., Banci, L., Bertini, I., Rosato, A., 2006. Counting the zinc-proteins encoded in the human genome. J Proteome Res 5 (1), 196–201.
Armony-Sivan, R., Eidelman, A.I., Lanir, A., Sredni, D., Yehuda, S., 2004. Iron status and neurobehavioral development of premature infants. J Perinatol 24 (12), 757–762.
Arons, M.H., Lee, K., Thynne, C.J., Kim, S.A., Schob, C., Kindler, S., Montgomery, J.M., Garner, C.C., 2016. Shank3 is part of a zinc-sensitive signaling system that regulates excitatory synaptic strength. J Neurosci 36 (35), 9124–9134.
Assaf, S.Y., Chung, S.H., 1984. Release of endogenous Zn^{2+} from brain tissue during activity. Nature 308 (5961), 734–736.
Bakhti, M., Aggarwal, S., Simons, M., 2014. Myelin architecture: zippering membranes tightly together. Cell Mol Life Sci 71 (7), 1265–1277.
Barceloux, D.G., 1999. Cobalt. J Toxicol Clin Toxicol 37, 201–206.
Baron, M.K., Boeckers, T.M., Vaida, B., Faham, S., Gingery, M., Sawaya, M.R., Salyer, D., Gundelfinger, E.D., Bowie, J.U., 2006. An architectural framework that may lie at the core of the postsynaptic density. Science 311 (5760), 531–535.
Beard, J., 2003. Iron deficiency alters brain development and functioning. J Nutr 133 (5 Suppl. 1), 1468–1472.
Bertoglio, K., Jill James, S., Deprey, L., Brule, N., Hendren, R.L., 2010. Pilot study of the effect of methyl B_{12} treatment on behavioral and biomarker measures in children with autism. J Altern Complement Med 16 (5), 555–560.
Bourgeron, T., 2009. A synaptic trek to autism. Curr Opin Neurobiol 19 (2), 231–234.
Briani, C., Dalla Torre, C., Citton, V., Manara, R., Pompanin, S., Binotto, G., Adami, F., 2013. Cobalamin deficiency: clinical picture and radiological findings. Nutrients 5 (11), 4521–4539.
Brown, C.E., Dyck, R.H., 2004. Distribution of zincergic neurons in the mouse forebrain. J Comp Neurol 479 (2), 156–167.
Chowanadisai, W., Graham, D.M., Keen, C.L., Rucker, R.B., Messerli, M.A., 2013. A zinc transporter gene required for development of the nervous system. Commun Integr Biol 6 (6), e26207.
Cole, T.B., Wenzel, H.J., Kafer, K.E., Schwartzkroin, P.A., Palmiter, R.D., 1999. Elimination of zinc from synaptic vesicles in the intact mouse brain by disruption of the ZnT_3 gene. Proc Natl Acad Sci USA 96 (4), 1716–1721.
Cousins, R.J., 1998. A role of zinc in the regulation of gene expression. Proc Nutr Soc 57 (2), 307–311.
Deak, F., Sonntag, W.E., 2012. Aging, synaptic dysfunction, and insulin-like growth factor (IGF)-1. J Gerontol A Biol Sci Med Sci 67 (6), 611–625.
DeLoughery, T.G., 2017. Iron deficiency anemia. Med Clin North Am 101 (2), 319–332.
Dipaolo, R.V., Kanfer, J.N., Newberne, P.M., 1974. Copper deficiency and the central nervous system. Myelination in the rat: morphological and biochemical studies. J Neuropathol Exp Neurol 33 (2), 226–236.
Dvergsten, C.L., Fosmire, G.J., Ollerich, D.A., Sandstead, H.H., 1983. Alterations in the postnatal development of the cerebellar cortex due to zinc deficiency. I. Impaired acquisition of granule cells. Brain Res 271 (2), 217–226.
Dvergsten, C.L., Johnson, L.A., Sandstead, H.H., 1984. Alterations in the postnatal development of the cerebellar cortex due to zinc deficiency. III. Impaired dendritic differentiation of basket and stellate cells. Brain Res 318 (1), 21–26.
Dyer, A.H., Vahdatpour, C., Sanfeliu, A., Tropea, D., 2016. The role of Insulin-Like Growth Factor 1 (IGF-1) in brain development, maturation and neuroplasticity. Neuroscience 325, 89–99.

Eng, L., Lam, L., 2020. Thyroid function during the fetal and neonatal periods. Neoreviews 21 (1), e30–e36.
Frazzini, V., Granzotto, A., Bomba, M., Massetti, N., Castelli, V., d'Aurora, M., Punzi, M., Iorio, M., Mosca, A., Delli Pizzi, S., Gatta, V., Cimini, A., Sensi, S.L., 2018. The pharmacological perturbation of brain zinc impairs BDNF-related signaling and the cognitive performances of young mice. Sci Rep 8 (1), 9768.
Frederickson, C.J., Koh, J.Y., Bush, A.I., 2005. The neurobiology of zinc in health and disease. Nat Rev Neurosci 6 (6), 449–462.
Gao, H.L., Zheng, W., Xin, N., Chi, Z.H., Wang, Z.Y., Chen, J., Wang, Z.Y., 2009. Zinc deficiency reduces neurogenesis accompanied by neuronal apoptosis through caspase-dependent and -independent signaling pathways. Neurotox Res 16 (4), 416–425.
Gaskin, F., Kress, Y., 1977. Zinc ion-induced assembly of tubulin. J Biol Chem 252 (19), 6918–6924.
Gonçalves, J.T., Schafer, S.T., Gage, F.H., 2016. Adult neurogenesis in the Hippocampus: from stem cells to behavior. Cell 167 (4), 897–914.
Gower-Winter, S.D., Corniola, R.S., Morgan, T.J., Levenson, C.W., 2013. Zinc deficiency regulates hippocampal gene expression and impairs neuronal differentiation. Nutr Neurosci 16 (4), 174–182.
Gower-Winter, S.D., Levenson, C.W., 2012. Zinc in the central nervous system: from molecules to behavior. Biofactors 38 (3), 186–193.
Grabrucker, A.M., 2014. A role for synaptic zinc in ProSAP/Shank PSD scaffold malformation in autism spectrum disorders. Dev Neurobiol 74 (2), 136–146.
Grabrucker, A.M., 2016. Zinc in the developing brain. In: Moran, V.H., Lowe, N.M. (Eds.), Nutrition and the Developing Brain. CRC Press, pp. 144–168.
Grabrucker, A.M., Knight, M.J., Proepper, C., Bockmann, J., Joubert, M., Rowan, M., Nienhaus, G.U., Garner, C.C., Bowie, J.U., Kreutz, M.R., Gundelfinger, E.D., Boeckers, T.M., 2011. Concerted action of zinc and ProSAP/Shank in synaptogenesis and synapse maturation. EMBO J 30 (3), 569–581.
Grabrucker, S., Jannetti, L., Eckert, M., Gaub, S., Chhabra, R., Pfaender, S., Mangus, K., Reddy, P.P., Rankovic, V., Schmeisser, M.J., Kreutz, M.R., Ehret, G., Boeckers, T.M., Grabrucker, A.M., 2014. Zinc deficiency dysregulates the synaptic ProSAP/Shank scaffold and might contribute to autism spectrum disorders. Brain 137 (Pt 1), 137–152.
Grabrucker, S., Boeckers, T.M., Grabrucker, A.M., 2016. Gender dependent evaluation of autism like behavior in mice exposed to prenatal zinc deficiency. Front Behav Neurosci 10, 37.
Grabrucker, S., Haderspeck, J.C., Sauer, A.K., Kittelberger, N., Asoglu, H., Abaei, A., Rasche, V., Schön, M., Boeckers, T.M., Grabrucker, A.M., 2017. Brain lateralization in mice is associated with zinc signaling and altered in prenatal zinc deficient mice that display features of autism spectrum disorder. Front Mol Neurosci 10, 450.
Gybina, A.A., Tkac, I., Prohaska, J.R., 2009. Copper deficiency alters the neurochemical profile of developing rat brain. Nutr Neurosci 12 (3), 114–122.
Ha, H.T.T., Leal-Ortiz, S., Lalwani, K., Kiyonaka, S., Hamachi, I., Mysore, S.P., Montgomery, J.M., Garner, C.C., Huguenard, J.R., Kim, S.A., 2018. Shank and zinc mediate an AMPA receptor subunit switch in developing neurons. Front Mol Neurosci 11, 405.
Hagmeyer, S., Haderspeck, J.C., Grabrucker, A.M., 2014. Behavioral impairments in animal models for zinc deficiency. Front Behav Neurosci 8, 443.
Hara, A., Taketomi, T., 1986. Cerebral lipid and protein abnormalities in Menkes' steely-hair disease. Jpn J Exp Med 56 (6), 277–284.
Haug, F.M., 1967. Electron microscopical localization of the zinc in hippocampal mossy fibre synapses by a modified sulfide silver procedure. Histochemie 8 (4), 355–368.

Hayward, L.J., Rodriguez, J.A., Kim, J.W., Tiwari, A., Goto, J.J., Cabelli, D.E., Valentine, J.S., Brown, R.H., 2002. Decreased metallation and activity in subsets of mutant superoxide dismutases associated with familial amyotrophic lateral sclerosis. J Biol Chem 277 (18), 15923–15931.

Heiliger, E., Osmanagic, A., Haase, H., Golenhofen, N., Grabrucker, A.M., Weth, A., Baumgartner, W., 2015. N-Cadherin-mediated cell adhesion is regulated by extracellular Zn(2+). Metallomics 7 (2), 355–362.

Hendren, R.L., James, S.J., Widjaja, F., Lawton, B., Rosenblatt, A., Bent, S., 2016. Randomized, placebo-controlled trial of methyl B_{12} for children with autism. J Child Adolesc Psychopharmacol 26 (9), 774–783.

Hershfinkel, M., 2018. The zinc sensing receptor, ZnR/GPR39, in health and disease. Int J Mol Sci 19 (2).

Hertz, L., Dringen, R., Schousboe, A., Robinson, S.R., 1999. Astrocytes: glutamate producers for neurons. J Neurosci Res 57 (4), 417–428.

Hesketh, J.E., 1982. Zinc-stimulated microtubule assembly and evidence for zinc binding to tubulin. Int J Biochem 14 (11), 983–990.

Hidalgo, J., Aschner, M., Zatta, P., Vasák, M., 2001. Roles of the metallothionein family of proteins in the central nervous system. Brain Res Bull 55 (2), 133–145.

Höck, A., Demmel, U., Schicha, H., Kasperek, K., Feinendegen, L.E., 1975. Trace element concentration in human brain. Activation analysis of cobalt, iron, rubidium, selenium, zinc, chromium, silver, cesium, antimony and scandium. Brain 98 (1), 49–64.

Holm, I.E., Andreasen, A., Danscher, G., Pérez-Clausell, J., Nielsen, H., 1988. Quantification of vesicular zinc in the rat brain. Histochemistry 89 (3), 289–293.

Horning, K.J., Caito, S.W., Tipps, K.G., Bowman, A.B., Aschner, M., 2015. Manganese is essential for neuronal health. Annu Rev Nutr 35, 71–108.

Howell, G.A., Welch, M.G., Frederickson, C.J., 1984. Stimulation-induced uptake and release of zinc in hippocampal slices. Nature 308 (5961), 736–738.

Hutto, B.R., 1997. Folate and cobalamin in psychiatric illness. Compr Psychiatry 38 (6), 305–314.

Insel, B.J., Schaefer, C.A., McKeague, I.W., Susser, E.S., Brown, A.S., 2008. Maternal iron deficiency and the risk of schizophrenia in offspring. Arch Gen Psychiatry 65 (10), 1136–1144.

Jan, H.H., Chen, I.T., Tsai, Y.Y., Chang, Y.C., 2002. Structural role of zinc ions bound to postsynaptic densities. J Neurochem 83 (3), 525–534.

Kalappa, B.I., Anderson, C.T., Goldberg, J.M., Lippard, S.J., Tzounopoulos, T., 2015. AMPA receptor inhibition by synaptically released zinc. Proc Natl Acad Sci USA 112 (51), 15749–15754.

Koh, J.Y., Lim, J.S., Byun, H.R., Yoo, M.H., 2014. Abnormalities in the zinc-metalloprotease-BDNF axis may contribute to megalencephaly and cortical hyperconnectivity in young autism spectrum disorder patients. Mol Brain 7, 64.

Leblond, C.S., Nava, C., Polge, A., Gauthier, J., Huguet, G., Lumbroso, S., Giuliano, F., Stordeur, C., Depienne, C., Mouzat, K., Pinto, D., Howe, J., Lemière, N., Durand, C.M., Guibert, J., Ey, E., Toro, R., Peyre, H., Mathieu, A., Amsellem, F., Rastam, M., Gillberg, I.C., Rappold, G.A., Holt, R., Monaco, A.P., Maestrini, E., Galan, P., Heron, D., Jacquette, A., Afenjar, A., Rastetter, A., Brice, A., Devillard, F., Assouline, B., Laffargue, F., Lespinasse, J., Chiesa, J., Rivier, F., Bonneau, D., Regnault, B., Zelenika, D., Delepine, M., Lathrop, M., Sanlaville, D., Schluth-Bolard, C., Edery, P., Perrin, L., Tabet, A.C., Schmeisser, M.J., Boeckers, T.M., Coleman, M., Sato, D., Szatmari, P., Scherer, S.W., Rouleau, G.A., Betancur, C., Leboyer, M., Gillberg, C., Delorme, R., Bourgeron, T., 2014. Meta-analysis of SHANK mutations in autism spectrum disorders: a gradient of severity in cognitive impairments. PLoS Genet 10 (9), e1004580.

Li, Y.V., Frederickson, C.J., 2013. Zinc-secreting neurons, gluzincergic and zincergic neurons. In: Kretsinger, R.H., Uversky, V.N., Permyakov, E.A. (Eds.), Encyclopedia of Metalloproteins. Springer, New York, NY.
Li, Y., Kim, J., Buckett, P.D., Böhlke, M., Maher, T.J., Wessling-Resnick, M., 2011. Severe postnatal iron deficiency alters emotional behavior and dopamine levels in the prefrontal cortex of young male rats. J Nutr 141 (12), 2133–2138.
Linder, M.C., Wooten, L., Cerveza, P., Cotton, S., Shulze, R., Lomeli, N., 1998. Copper transport. Am J Clin Nutr 67 (5 Suppl.), 965–971.
Liu, H., Oteiza, P.I., Gershwin, M.E., Golub, M.S., Keen, C.L., 1992. Effects of maternal marginal zinc deficiency on myelin protein profiles in the suckling rat and infant rhesus monkey. Biol Trace Elem Res 34 (1), 55–66.
Liu, M.J., Bao, S., Gálvez-Peralta, M., Pyle, C.J., Rudawsky, A.C., Pavlovicz, R.E., Killilea, D.W., Li, C., Nebert, D.W., Wewers, M.D., Knoell, D.L., 2013. ZIP8 regulates host defense through zinc-mediated inhibition of NF-κB. Cell Rep 3 (2), 386–400.
Lozoff, B., Jimenez, E., Hagen, J., Mollen, E., Wolf, A.W., 2000. Poorer behavioral and developmental outcome more than 10 years after treatment for iron deficiency in infancy. Pediatrics 105 (4), E51.
Lozoff, B., Klein, N.K., Nelson, E.C., McClish, D.K., Manuel, M., Chacon, M.E., 1998. Behavior of infants with iron-deficiency anemia. Child Dev 69 (1), 24–36.
Lukowski, A.F., Koss, M., Burden, M.J., Jonides, J., Nelson, C.A., Kaciroti, N., Jimenez, E., Lozoff, B., 2010. Iron deficiency in infancy and neurocognitive functioning at 19 years: evidence of long-term deficits in executive function and recognition memory. Nutr Neurosci 13 (2), 54–70.
Maret, W., 2017. Zinc in cellular regulation: the nature and significance of "zinc signals". Int J Mol Sci 18 (11).
McCall, K.A., Huang, C., Fierke, C.A., 2000. Function and mechanism of zinc metalloenzymes. J Nutr 130 (5S Suppl.), 1437–1446.
MacDonald, R.S., 2000. The role of zinc in growth and cell proliferation. J Nutr 130 (5S Suppl.), 1500–1508.
Mendel, R.R., Bittner, F., 2006. Cell biology of molybdenum. Biochim Biophys Acta 1763 (7), 621–635.
Miller, A.L., 2008. The methylation, neurotransmitter, and antioxidant connections between folate and depression. Altern Med Rev 13 (3), 216–226.
Montes, S., Alcaraz-Zubeldia, M., Muriel, P., Rios, C., 2003. Role of manganese accumulation in increased brain glutamine of the cirrhotic rat. Neurochem Res 28 (6), 911–917.
Morath, D.J., Mayer-Pröschel, M., 2001. Iron modulates the differentiation of a distinct population of glial precursor cells into oligodendrocytes. Dev Biol 237 (1), 232–243.
Mott, D.D., Benveniste, M., Dingledine, R.J., 2008. pH-dependent inhibition of kainate receptors by zinc. J Neurosci 28 (7), 1659–1671.
Muñoz, P., Humeres, A., 2012. Iron deficiency on neuronal function. Biometals 25 (4), 825–835.
Nieto-Estévez, V., Defterali, Ç., Vicario-Abejón, C., 2016. IGF-I: a key growth factor that regulates neurogenesis and synaptogenesis from embryonic to adult stages of the brain. Front Neurosci 10, 52.
Ortiz, E., Pasquini, J.M., Thompson, K., Felt, B., Butkus, G., Beard, J., Connor, J.R., 2004. Effect of manipulation of iron storage, transport, or availability on myelin composition and brain iron content in three different animal models. J Neurosci Res 77 (5), 681–689.
Oteiza, P.I., Cuellar, S., Lönnerdal, B., Hurley, L.S., Keen, C.L., 1990. Influence of maternal dietary zinc intake on in vitro tubulin polymerization in fetal rat brain. Teratology 41 (1), 97–104.

Palmiter, R.D., Cole, T.B., Quaife, C.J., Findley, S.D., 1996. ZnT-3, a putative transporter of zinc into synaptic vesicles. Proc Natl Acad Sci USA 93 (25), 14934–14939.

Pang, W., Leng, X., Lu, H., Yang, H., Song, N., Tan, L., Jiang, Y., Guo, C., 2013. Depletion of intracellular zinc induces apoptosis of cultured hippocampal neurons through suppression of ERK signaling pathway and activation of caspase-3. Neurosci Lett 552, 140–145.

Peirano, P.D., Algarín, C.R., Chamorro, R.A., Reyes, S.C., Durán, S.A., Garrido, M.I., Lozoff, B., 2010. Sleep alterations and iron deficiency anemia in infancy. Sleep Med 11 (7), 637–642.

Pfaender, S., Föhr, K., Lutz, A.K., Putz, S., Achberger, K., Linta, L., Liebau, S., Boeckers, T.M., Grabrucker, A.M., 2016. Cellular zinc homeostasis contributes to neuronal differentiation in human induced pluripotent stem cells. Neural Plast 2016, 3760702.

Pivina, L., Semenova, Y., Dosa, M.D., Dauletyarova, M., Bjørklund, G., 2019. Iron deficiency, cognitive functions, and neurobehavioral disorders in children. J Mol Neurosci 68 (1), 1–10.

Pochwat, B., Nowak, G., Szewczyk, B., 2015. Relationship between zinc (Zn(2+)) and glutamate receptors in the processes underlying neurodegeneration. Neural Plast 2015, 591563.

Raghavan, R., Riley, A.W., Volk, H., Caruso, D., Hironaka, L., Sices, L., Hong, X., Wang, G., Ji, Y., Brucato, M., Wahl, A., Stivers, T., Pearson, C., Zuckerman, B., Stuart, E.A., Landa, R., Fallin, M.D., Wang, X., 2018. Maternal multivitamin intake, plasma folate and vitamin B_{12} levels and autism spectrum disorder risk in offspring. Paediatr Perinat Epidemiol 32 (1), 100–111.

Rehnberg, G.L., Hein, J.F., Carter, S.D., Linko, R.S., Laskey, J.W., 1982. Chronic ingestion of Mn_3O_4 by rats: tissue accumulation and distribution of manganese in two generations. J Toxicol Environ Health 9, 175–188.

Rockland, K.S., 2014. Zinc-positive and zinc-negative connections of the claustrum. Front Syst Neurosci 8, 37.

Robinett, N.G., Peterson, R.L., Culotta, V.C., 2018. Eukaryotic copper-only superoxide dismutases (SODs): a new class of SOD enzymes and SOD-like protein domains. J Biol Chem 293 (13), 4636–4643.

Sanford, L., Carpenter, M.C., Palmer, A.E., 2019. Intracellular Zn^{2+} transients modulate global gene expression in dissociated rat hippocampal neurons. Sci Rep 9 (1), 9411.

Sardesai, V.M., 1993. Molybdenum: an essential trace element. Nutr Clin Pract 8 (6), 277–281.

Scalabrino, G., 2009. The multi-faceted basis of vitamin B-12 (cobalamin) neurotrophism in adult central nervous system: lessons learned from its deficiency. Prog Neurobiol 88, 203–220.

Schmidt, K., Ralle, M., Schaffer, T., Jayakanthan, S., Bari, B., Muchenditsi, A., Lutsenko, S., 2018. ATP7A and ATP7B copper transporters have distinct functions in the regulation of neuronal dopamine-β-hydroxylase. J Biol Chem 293 (52), 20085–20098.

Seth, R., Corniola, R.S., Gower-Winter, S.D., Morgan, T.J., Bishop, B., Levenson, C.W., 2015. Zinc deficiency induces apoptosis via mitochondrial p53- and caspase-dependent pathways in human neuronal precursor cells. J Trace Elem Med Biol 30, 59–65.

Sidoryk-Wegrzynowicz, M., Aschner, M., 2013. Role of astrocytes in manganese mediated neurotoxicity. BMC Pharmacol Toxicol 14, 23.

Steffens, G.C., Biewald, R., Buse, G., 1987. Cytochrome c oxidase is a three-copper, two-heme-A protein. Eur J Biochem 164 (2), 295–300.

Steinman, G., 2015. Plausible etiology of brain dysconnectivity in autism - review and prospectus. Med Hypotheses 85 (4), 405–407.

Steinman, G., 2019. IGF - autism prevention/amelioration. Med Hypotheses 122, 45–47.
Suh, S.W., Won, S.J., Hamby, A.M., Yoo, B.H., Fan, Y., Sheline, C.T., Tamano, H., Takeda, A., Liu, J., 2009. Decreased brain zinc availability reduces hippocampal neurogenesis in mice and rats. J Cerebr Blood Flow Metabol 29 (9), 1579–1588.
Takeda, A., 2003. Manganese action in brain function. Brain Res Brain Res Rev 41 (1), 79–87.
Takeda, A., Sotogaku, N., Oku, N., 2002. Manganese influences the levels of neurotransmitters in synapses in rat brain. Neuroscience 114 (3), 669–674.
Takeda, A., Takada, S., Ando, M., Itagaki, K., Tamano, H., Suzuki, M., Iwaki, H., Oku, N., 2010. Impairment of recognition memory and hippocampal long-term potentiation after acute exposure to clioquinol. Neuroscience 171 (2), 443–450.
Tamura, T., Goldenberg, R.L., Hou, J., Johnston, K.E., Cliver, S.P., Ramey, S.L., Nelson, K.G., 2002. Cord serum ferritin concentrations and mental and psychomotor development of children at five years of age. J Pediatr 140 (2), 165–170.
Tao-Cheng, J.H., Toy, D., Winters, C.A., Reese, T.S., Dosemeci, A., 2016. Zinc stabilizes Shank3 at the postsynaptic density of hippocampal synapses. PLoS One 11 (5), e0153979.
Tran, P.V., Fretham, S.J., Wobken, J., Miller, B.S., Georgieff, M.K., 2012. Gestational-neonatal iron deficiency suppresses and iron treatment reactivates IGF signaling in developing rat hippocampus. Am J Physiol Endocrinol Metab 302 (3), E316–E324.
Tsang, D., Tsang, Y.S., Ho, W.K., Wong, R.N., 1997. Myelin basic protein is a zinc-binding protein in brain: possible role in myelin compaction. Neurochem Res 22 (7), 811–819.
Upadhyay, M., Agarwal, S., 2020. Ironing the mitochondria: relevance to its dynamics. Mitochondrion 50, 82–87.
Van Rij, A.M., Thomson, C.D., McKenzie, J.M., Robinson, M.F., 1979. Selenium deficiency in total parenteral nutrition. Am J Clin Nutr 32, 2076–2085.
Vergnano, A.M., Rebola, N., Savtchenko, L.P., Pinheiro, P.S., Casado, M., Kieffer, B.L., Rusakov, D.A., Mulle, C., Paoletti, P., 2014. Zinc dynamics and action at excitatory synapses. Neuron 82 (5), 1101–1114.
Wachs, T.D., Pollitt, E., Cueto, S., Jacoby, E., Creed-Kanashiro, H., 2005. Relation of neonatal iron status to individual variability in neonatal temperament. Dev Psychobiol 46 (2), 141–153.
Wang, F.D., Bian, W., Kong, L.W., Zhao, F.J., Guo, J.S., Jing, N.H., 2001. Maternal zinc deficiency impairs brain nestin expression in prenatal and postnatal mice. Cell Res 11 (2), 135–141.
Wolf, C., Weth, A., Walcher, S., Lax, C., Baumgartner, W., 2018. Modeling of zinc dynamics in the synaptic cleft: implications for cadherin mediated adhesion and synaptic plasticity. Front Mol Neurosci 11, 306.
Wrigley, S., Arafa, D., Tropea, D., 2017. Insulin-like growth factor 1: at the crossroads of brain development and aging. Front Cell Neurosci 11, 14.
Zhang, Y., Hodgson, N.W., Trivedi, M.S., Abdolmaleky, H.M., Fournier, M., Cuenod, M., Do, K.Q., Deth, R.C., 2016. Decreased brain levels of vitamin B_{12} in aging, autism and schizophrenia. PLoS One 11 (1), e0146797.

CHAPTER 5

Nonessential metals and their brain pathology

Contents

Pathology of toxic metals

Like essential trace metals, toxic metals may affect healthy brain development and function in various ways including epigenetic mechanisms (e.g., DNA methylation, histone modifications, or expression of microRNAs), altering protein function, or influencing gut–brain signaling through changing the microbiota composition (Zhai et al., 2019). Regarding autism spectrum disorder (ASD), the impacts may be limited to an individual's prenatal period, although the effects such as the alteration in gene expression may only appear later in life (Dietert et al., 2011). Given that the placenta does not provide effective protection against the transfer of many toxic metals to the embryo (Chen et al., 2014) and that the blood–brain barrier (BBB) of the embryo is only forming, the phase of early development is a significant window of vulnerability (Rodier, 1995).

Biometals in Autism Spectrum Disorders
ISBN 978-0-12-821132-8
https://doi.org/10.1016/B978-0-12-821132-8.00005-7

Neurotoxicity of lead

Prenatal lead exposure may not only be due to acute maternal exposure but may also be caused by mobilization of lead accumulated in the mother's bones through past exposures. Lead was shown to cross the placenta resulting in a blood lead concentration of babies that is approximately 90% of that of the mother (Amitai et al., 1999).

The effects of lead in the central nervous system (CNS) include increased neuronal death, impaired neuronal proliferation, migration, differentiation, synaptogenesis and synaptic plasticity, myelinization, and increased inflammatory cytokine release (Coria et al., 1984; Toscano and Guilarte, 2005; Lidsky and Schneider, 2006; Chibowska et al., 2016). On the gross morphological level, lead exposure has been linked to volume changes and altered connectivity of the brain (Cecil et al., 2008; Thomason et al., 2019). However, how lead induces these effects on the molecular level is still not completely understood.

One described pathomechanism of lead is the disruption of physiological processes mediated by essential metals such as calcium and zinc (Bressler and Goldstein, 1991). Lead causes damage to the mitochondria, which results in calcium release from mitochondria (Kerper and Hinke, 1997). Calcium depletion of mitochondria may trigger apoptosis. In addition, decreased mitochondrial functioning can increase excitotoxicity. However, lead not only interferes with mitochondrial calcium but also substitutes calcium in calcium signaling processes, thereby, for example, affecting calmodulin and several downstream protein kinases (Bressler et al., 1999). Thereby, lead impairs synaptic signaling and essential cellular functions, including proliferation and differentiation, for example, through altering protein kinase C (PKC) activity. Besides, lead decreases the activity and calcium-dependent release of neurotransmitters vesicles (Lasley et al., 1999).

Lead can also displace zinc. Thus, lead overload may mimic some of the effects of zinc deficiency. Lead may even affect high-affinity zinc binding that occurs in zinc finger proteins, most of which are involved in the regulation of gene expression.

Furthermore, lead can affect lipid metabolism (Shukla et al., 1987; Villeda-Hernández et al., 2001). Besides causing neuronal cell death through excitotoxicity and mitochondrial damage, oxidative stress, either directly through elevated lead levels or indirectly through increased lipid peroxidation, may trigger apoptosis.

In addition to affecting calcium-dependent neurotransmitter release, lead can impair mechanisms of synapse formation and plasticity by altering neurotransmitter receptor function. One primary target of lead is glutamate receptors, in particular, NMDA and AMPA receptors (McCoy et al., 1997; Lasley et al., 2001).

Moreover, lead accumulates in glial cells such as astrocytes and oligodendrocytes (Lindahl et al., 1999; Lidsky and Schneider, 2003). Impaired oligodendrocyte function may explain impaired myelinization, for example, through a decrease in the activity of CNPase, an enzyme shown to be critical for myelin synthesis (Dabrowska-Bouta et al., 2000).

Another effect of lead on the brain involves the disruption of thyroid hormone transport into the CNS (Lidsky and Schneider, 2003). Thyroid hormones are critical during brain development, acting as mediators of neural cell migration and differentiation, synaptogenesis, and myelination (Bernal, 2007).

Neurotoxicity of mercury

After inhalation in the form of mercury vapor and subsequent adsorption in the lung epithelial cells, or adsorption in the gastrointestinal tract, mercury may reach the CNS due to its ability to cross the BBB (Aschner and Aschner, 1990). For example, methylmercury (MeHg) crosses the BBB mainly using the L-type neutral amino acid carrier transport (LAT) system, and in particular, the LAT1 transporter, on endothelial cells. However, it has been reported that mercury exposure also results in damage to the BBB, which will result in the exposure of the CNS to further unwanted elements that will trigger processes such as neuroinflammation (Takahashi and Shimohata, 2019). During pregnancy, mercury is also able to cross the blood—placenta barrier (Kern et al., 2020; Sakamoto et al., 2018).

Inside the brain, mercury can be found extra- as well as intracellular (Cariccio et al., 2019). Once inside cells, mercury increases the levels of oxidative stress and damages mitochondria (Yin et al., 2007; Farina and Aschner, 2019; Carocci et al., 2014), resulting, among others, in high lipid peroxidation products that cause selective alterations in cell signaling, protein and DNA damage, and cytotoxicity, and impairment of energy metabolism (Ramana et al., 2013; Morris et al., 2018).

Within the CNS, mercury seems to get enriched in astrocytes and, to a lesser extent, in microglial cells (Shanker et al., 2003; Tiffany-Castiglion and Qian, 2001), leading to dysfunction of these cells (Pieper et al., 2014). However, neurons seem to be more susceptible to mercury-induced toxicity (Farina et al., 2011; Guzzi and La Porta, 2008). A significant reduction in the number of neurons as a consequence of high mercury levels has been described (Carvalho et al., 2008; Falluel-Morel et al., 2007).

Several molecular mechanisms have been discussed for the effects of mercury and in particular MeHg, on neurons. Mercury combines covalently with sulfhydryl (thiol) groups, which can result in the inhibition of sulfhydryl-containing enzymes (Farina et al., 2011). Similarly, the direct chemical interaction between MeHg and thiol may affect other proteins and nonprotein molecules, such as glutathione (GSH) (Farina et al., 2011). As a consequence, GSH levels were shown to decrease, which leads to an increase in oxidative stress (Franco et al., 2009) due to the role of GSH as an antioxidant. However, MeHg was also shown to induce oxidative stress due to direct interaction with nucleophilic protein groups (Farina et al., 2011).

Apart from neuronal death caused by oxidative stress, the presence of mercury in the CNS also leads to glutamate excitotoxicity. Glutamate is the major excitatory neurotransmitter. In vitro and animal studies have shown that MeHg inhibits the reuptake of glutamate into synaptic vesicles (Porciúncula et al., 2003; Moretto et al., 2005), thereby increasing glutamate levels in the synaptic cleft (Farina et al., 2003). In addition, MeHg also increases the spontaneous release of glutamate (Reynolds and Racz, 1987; Vendrell et al., 2007). The underlying processes may be triggered by an increase in hydrogen peroxide levels as a response to MeHg, acting on glutamate transporters, or the direct inhibition of H^{+}-ATPase activity (Porciúncula et al., 2003).

Abnormally high concentrations of glutamate in the synaptic cleft can induce neuronal damage and death (Niciu et al., 2012). As a consequence, *N*-methyl D-aspartate (NMDA)—type glutamate receptors become overactivated, leading to increased sodium and calcium influx into neurons. Excessive intracellular calcium, in return, will again increase oxidative stress (e.g., through uptake into mitochondria) and interfere with the activation of important intracellular calcium-dependent signaling pathways. Since astrocytes contribute to glutamatergic signaling by removing the neurotransmitter from the synaptic cleft, damage of astrocytes by mercury may further exacerbate excitotoxicity (Aschner et al., 2000).

Mercury is mostly cleared from the body through binding to GSH in bile and subsequent excretion in feces (Ballatori and Clarkson, 1984) as Hg-GSH complex. It was found that children with ASD have significantly lower plasma concentrations of total glutathione (tGSH) but significantly higher concentrations of oxidized glutathione (GSSG) (James et al., 2004).

Neurotoxicity of cadmium

The molecular mechanisms of cadmium toxicity in the CNS are still not well investigated. However, some effects of cadmium that directly or indirectly may impair brain development have been uncovered. For example, high levels of cadmium may cause oxidative stress through indirect mechanisms: Cadmium has an affinity for thiol groups, forming cadmium—thiol complexes inside cells (Figueiredo-Pereira et al., 1998). As a consequence, glutathione levels decrease, which affects cellular activities against reactive oxygen species (ROS) (Bertin and Averbeck, 2006). Also, cadmium disturbs the cellular protein degradation process, which becomes evident through increased levels of ubiquitinated proteins. As a consequence, cadmium interferes with cellular processes such as cell cycle control and cell differentiation and may induce apoptosis (Dong et al., 2001; Yang et al., 2004).

Moreover, elevated cadmium interferes with the balance of essential trace metals. In particular, imbalances in zinc transport to the fetus, causing deficiency, have been reported (Kippler et al., 2010). Therefore, the effects of zinc deficiency apply to some extent to cadmium toxicity. Furthermore, developmental cadmium toxicity may affect the expression of thyroid hormones (Ishitobi et al., 2007), whose transport was affected by lead.

Pathomechanisms of toxic metals in autism spectrum disorder

Many parallels between the effects of mercury, lead, and cadmium intoxication on the brain and the brain pathologies reported in individuals with ASD can be found. In general, the primary pathologies observed after exposure to high levels of toxic metals that may be active in the developing brain are (1) competition with essential metals, especially zinc, (2) oxidative stress and lipid peroxidation, (3) mitochondrial dysfunction, (4) neuroinflammation and gliosis, and (6) axonal demyelination (Geier et al., 2010; Kern et al., 2012) (Fig. 5.1).

Competition with essential metals

Toxic metals may interfere with processes mediated by essential metals through competing with them for protein binding. For example, zinc and lead compete for similar binding sites on enzymes (Flora et al., 1989, 1991). This may be one mechanism of how lead interferes with hippocampal glutamatergic neurotransmission (Neal et al., 2010). Lead may also compete with calcium by altering its transport (Flora et al., 2007).

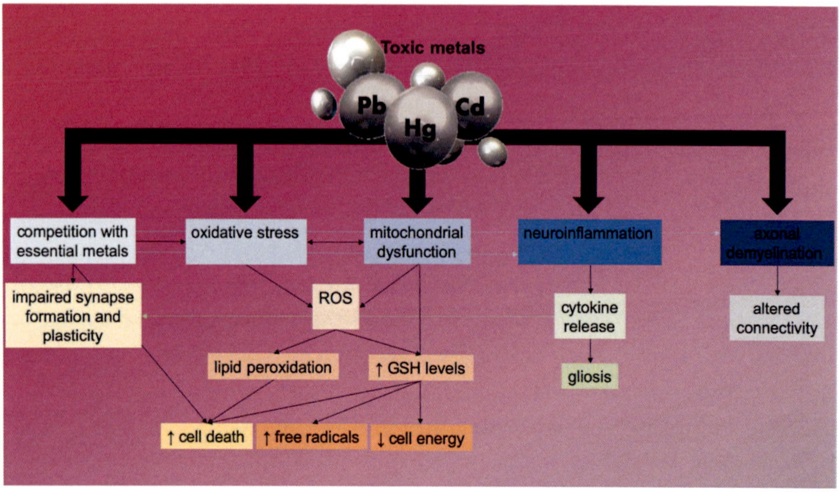

Figure 5.1 ***Effects of toxic trace metals:*** Common features of high levels of mercury, lead, and cadmium are the competition with essential trace metals such as zinc that has several important roles in the brain such as mediating neurogenesis and neural differentiation, synaptic function and maturation, myelination, as well as regulating inflammatory processes and acting as antioxidant. Besides, toxic metals increase oxidative stress and cause mitochondrial dysfunction, thereby leading to lipid peroxidation, low energy levels, and, ultimately, cell death. Toxic metals further act proinflammatory, e.g., by damaging the BBB or activating astrocytes and microglia, which may result in gliosis. They also affect oligodendrocytes and, thereby, axonal myelination processes.

Mercury was also found to interact with zinc. Zinc supplementation, therefore, reduces the effects of mercury (Franciscato et al., 2011) that may displace zinc from Zn metalloproteinases (Guzzi and La Porta, 2008).

Cadmium can compete with calcium for absorption in the gastrointestinal system (Jacobo-Estrada et al., 2017). Moreover, cadmium can compete with calcium, but also zinc transport in other tissues (Yang and Shu, 2015) and thereby affect the function of enzymes and second messengers, consequently disrupting cellular signaling, transport, and metabolism (Thévenod, 2009). In addition, cadmium should be competing with magnesium, copper, and iron to some extent (Moulis, 2010).

Oxidative stress and lipid peroxidation

Oxidative stress is a characteristic feature of ASD (Bjørklund et al., 2020a; Chauhan et al., 2012; Deth et al., 2008; Kern and Jones, 2006; Manivasagam et al., 2020). The production of free radicals or ROS is a natural by-product of the processes taking place in mitochondria that lead to the production of ATP. Besides, reactive nitrogen species (RNS) and carbon- and sulfur-centered radicals are generated (Salim, 2017). Therefore, the presence of low amounts of ROS in brain tissue should be considered normal, whereas excessive levels, resulting in oxidative stress, are damaging. Thus, oxidative stress may be either caused by excessive free radical production and/or insufficiency in the processes that occur in cells to reduce oxidative stress (the antioxidant response system).

The brain has high energy demands, and therefore, mitochondria need to generate high levels of ATP, which makes this organ particularly susceptible to oxidative stress. Besides, the brain consists of lipid-rich tissue. Thus, lipid peroxidation is especially problematic for neural cells (Salim, 2017). In addition to lipid peroxidation, protein oxidation occurs as a consequence of excess ROS. Together, the two processes are primarily responsible for neural cell death and functional impairments such as abnormal neurogenesis and synaptic signaling and plasticity (Knapp and Klann, 2002).

Currently, the molecular mechanisms of how excessive amounts of ROS induce cell death and functional impairments are not completely understood, but several studies indicate that an increase in BBB permeability causing neuroinflammation may be involved (Pun et al., 2009; Gu et al., 2011). Furthermore, abnormal glucocorticoid receptor signaling, excitotoxicity, and impaired *N*-methyl-D-aspartate (NMDA) receptor signaling have been implied (Kasahara and Inoue, 2015; Forder and Tymianski, 2009; Bórquez et al., 2016).

The antioxidant response system is a protective mechanism that can reduce the levels of ROS. It acts via two major processes: One is the activation of antioxidative enzymes such as superoxide dismutase (SOD), glyoxalase, glutathione reductase, glutathione peroxidase, and catalase (CAT) (Gemma et al., 2007). In the case of excess ROS, the transcription factor nuclear factor erythroid 2–related factor (Nrf2) is increasing the expression of many of the antioxidative enzymes (Lim et al., 2014). Intriguingly, SOD enzymes need essential trace metals for their function (e.g., copper and zinc: Cu–Zn SOD) that mediate spontaneous dismutation of superoxide radicals by SOD to generate H_2O_2, which is then removed by CAT and glutathione peroxidase enzymes. Therefore, toxic metals such as cadmium, lead, and mercury not only stimulate the generation of free radicals but also impair the antioxidative system by damaging antioxidative enzymes (e.g., by inhibiting functional SH groups) directly and by competing with essential trace metals necessary for enzyme function. The second process of the antioxidant response system involves low-molecular-weight antioxidants such as glutathione, uric acid, ascorbic acid, and melatonin. These function mainly by acting as a scavenger of metals (Birben et al., 2012).

Concerning the development of ASD, the induction of proinflammatory processes by increased oxidative stress may be a significant contributing factor (Ghanizadeh et al., 2012) (see the following). Neuroinflammation may be caused by increased BBB permeability. However, oxidative stress also affects nuclear factor-κB (NF-κB) (Lingappan, 2018). NF-κB is an important transcription factor that mediates inflammation and apoptosis and has been linked to ASD in several studies (Malik et al., 2011).

In addition, oxidative stress may alter the levels of brain-derived neurotrophic factor (BDNF). A link between BDNF and oxidative stress in ASD was observed (Ghanizadeh, 2012). BDNF is a member of the neurotrophin family of growth factors (Binder and Scharfman, 2004) and has been associated with the etiology of ASD as a modulator of neurogenesis, neuron survival, and synaptic plasticity (Saghazadeh and Rezaei, 2017).

Mitochondrial dysfunction

Mitochondria cellular organelles oxidize glucose and fatty acids to generate adenosine triphosphate (ATP). Besides energy production, mitochondria are also important mediators of programmed cell death (apoptosis) and calcium homeostasis. In neurons in the CNS, they influence processes such as synaptic plasticity and neurotransmitter release (Devine and Kittler, 2018). While some proteins necessary for mitochondrial function are encoded in their cyclic DNA found in mitochondria (mitochondrial DNA [mtDNA]), others are found encoded in the nuclear DNA. The number of mitochondria in each cell can vary. Cells with high energy demand, such as muscle, brain, and gastrointestinal cells, have a greater number of mitochondria.

Several studies have reported that mitochondrial dysfunction is associated with ASD (Weissman et al., 2008; Giulivi et al., 2010; Dhillon et al., 2011; Rossignol and Frye, 2012; Griffiths and Levy, 2017; Valiente-Pallejà et al., 2018; Castora, 2019). Mitochondrial damage is closely linked to an increase in oxidative stress, as mitochondria lack the enzymes necessary for the production of GSH. Therefore, they are particularly vulnerable to effects from free radicals (James et al., 2009). Mitochondrial damage in children with ASD manifests as mitochondrial dysfunction (lower mitochondrial-dependent oxygen consumption and higher hydrogen peroxide (H_2O_2) production), mtDNA overreplication, and mtDNA deletions (Giulivi et al., 2010; Varga et al., 2018). The increased H_2O_2 levels further contribute to oxidative stress in ASD. However, so far, it is not clear whether the mitochondrial dysfunction in children with ASD is a causative factor or occurs as a consequence of the same factors that lead to ASD, but with limited contribution to the ASD phenotype. Nevertheless, it was hypothesized that symptoms found as a consequence of mitochondrial dysfunction/mitochondrial disease overlap with symptoms of ASD, and therefore, a contribution of mitochondrial damage to the ASD phenotype may be likely (Frye and Rossignol, 2011). For example, developmental delay and developmental regression (i.e., loss of skills) occur in ASD but also in children with mitochondrial disease (Frye and Rossignol, 2011). It has been speculated that reduced synaptic neurotransmitter release due to mitochondrial dysfunction may be a possible pathomechanism. Especially very active neurons, such as inhibitory interneurons (gamma-aminobutyric acid [GABA]ergic neurons) (Segerstråle et al., 2010) will be particularly affected, leading to an imbalance between excitation and inhibition that has been implicated in the pathogenesis of ASD (Rubenstein and Merzenich, 2003). Furthermore, abnormal calcium signaling has also been implicated in ASD (Nguyen et al., 2018). Mitochondria act as calcium stores and play a critical role in the regulation of plasma membrane calcium channels (Napolioni et al., 2011). Also, exercise intolerance (the inability or decreased ability to perform physical exercise) has been reported in ASD and is also a prominent feature of mitochondrial disease (Weissmann et al., 2008). Heart and kidney defects are also observed in mitochondrial disease and are known comorbidities of some syndromic forms of ASD (Finsterer and Scorza, 2017; Limongelli et al., 2017). However, so far, the overlap of symptoms of mitochondrial disease and ASD has not been well studied on a mechanistic level.

Neuroinflammation and gliosis

Exposure to toxic metals can result in a range of immune abnormalities. Inflammation and immune system abnormalities have been linked to ASD in epidemiological studies and mechanistic studies using animal models. The role of essential trace metals in antiinflammatory regulation has been discussed in Chapter 4. Toxic trace metals have been reported to cause inflammation through various mechanisms. For example, mercury, cadmium, and lead were shown to alter gene expression in in vitro and mouse models directly. Among the genes altered were those whose encoded proteins function in processes such as transport, oxidative stress, heat shock, angiogenesis, and inflammation (Breton et al., 2013; Bartosiewicz et al., 2001; Eyssen-Hernandez et al., 1996). Furthermore, lead and methylmercury were shown to directly induce glial reactivity (Monnet-Tschudi et al., 2006), and the increase in ROS and BBB permeability may be contributing factors to this. Reactive microglia and astrocytes release proinflammatory cytokines such as TNFα, IL-1β, and IL-6 (Vargas et al., 2005). In recent years, many studies have reported increased cytokine levels in individuals with ASD (Siniscalco et al., 2018). For example, increased serum expression of IL-1β and IL-6 were reported as a result of the Childhood Autism Risks from Genetics and Environment (CHARGE) study. The increase in cytokine levels correlated with impairments in stereotypical behaviors and regression. In addition, TNF-α was found to increase in the serum of individuals with ASD and was positively correlated with ASD severity (Siniscalco et al., 2018). More importantly, proinflammatory processes and increased levels of cytokines were also found in the CNS. In human postmortem samples of individuals with ASD, increased levels of IL-1β, IL-6, IL-17, and TNF-α were detected (Wei et al., 2013; Theoharides et al., 2016). Animal studies have shown that IL-6, in particular, can alter brain development and function by affecting synapse formation, synapse morphology, and plasticity (Wei et al., 2012a, 2012b). In addition, IL-6 plays a critical role in autoimmune regulation (Kany et al., 2019).

Mercury was also shown to trigger autoimmunity (Pamphlett and Kum Jew, 2019; Vas and Monestier, 2008). Several studies show an increased prevalence of familial autoimmune disorders in ASD (Hughes et al., 2018; Edmiston et al., 2017). Maternal autoimmunity was, therefore, proposed to play a role in ASD by influencing neurodevelopment through the gestational immune environment. In particular, maternal autoantibodies (IgG) that can cross the placenta were discussed in relation to ASD. IgGs normally are protective immune factors. However, maternal autoantibodies may target antigens in the developing fetal brain and thereby inhibit the function of targeted proteins or activate them. Besides, IgGs might trigger neuroinflammatory processes. A subset of mothers of children with ASD (10%—12%) has been found to harbor autoantibodies (Hughes et al., 2018). Preliminary studies using animal models suggest that some of these antibodies may indeed be linked to an ASD-like pathology as causative factors.

Several target proteins of autoantibodies have been identified. For example, lactate dehydrogenases A and B (LDH-A, LDH-B) that are part of the regulation of cellular energy metabolism, stress-induced phosphoprotein 1 (STIP1) that plays a role in neuritogenesis and mediates neuronal survival, guanine deaminase (GDA) that is involved in dendritic branching of hippocampal neurons, collapsin response mediator proteins 1 and 2 (CRMP1, CRMP2) that influence growth cone collapse, cell migration and axon—dendrite specification, and Y-box binding protein 1(YBX1) that participates in DNA- and mRNA-dependent processes (Edmiston et al., 2017).

Increased mercury and lead levels result in an increased expression of proinflammatory cytokines and autoantibody generation already at low chronic exposure, especially in susceptible individuals (Mishra, 2009; Pollard et al., 2019; Bjørklund et al., 2020b). One study, for example, found an association between elevated mercury levels and the upregulation of 3760 autoantibodies in blood serum (Motts et al., 2014). Among the target proteins of lead and mercury-induced autoantibodies are neuronal cytoskeletal proteins, neurofilaments (NFs), and myelin basic protein (MBP) (El-Fawal et al., 1999).

Together, these findings indicate that toxic metal exposure can trigger autoimmune dysfunction and systemic inflammation in humans that, during pregnancy, may affect embryonic brain development in particular.

Axonal demyelination

Several studies have implicated a white matter pathology such as hypomyelination in ASD, suggesting an impaired communication between different brain regions (Lazar et al., 2014; Shen et al., 2018). During myelination in the CNS, neuronal axons are insulated with myelin, a multilamellar membrane rich in lipids produced by oligodendrocytes. Toxic trace metals affect myelination (Maiuolo et al., 2019), thereby impairing electrical conduction. As underlying molecular mechanisms, imbalanced intracellular calcium ion regulation, altered lipid formation, and imbalanced myelin formation have been proposed (Maiuolo et al., 2019). However, the exact mechanisms have not yet been investigated. In addition, the competition of toxic metals with essential metals, in particular zinc, may contribute to abnormal myelination. In a study using rhesus monkeys, maternal zinc deficiency was shown to result in altered myelin protein profiles in the offspring (Liu et al., 1992). Further, myelin basic protein (MBP), a target of mercury and lead-induced autoantibodies, was also identified as zinc-binding protein establishing a role for zinc in myelin compaction (Tsang et al., 1997; Bakhti et al., 2014).

Conclusions

There are recurring pathological differences in individuals with ASD that are caused by genetic and nongenetic factors, including the lack of essential trace metals or the exposure to excessive amounts of toxic metals. Among

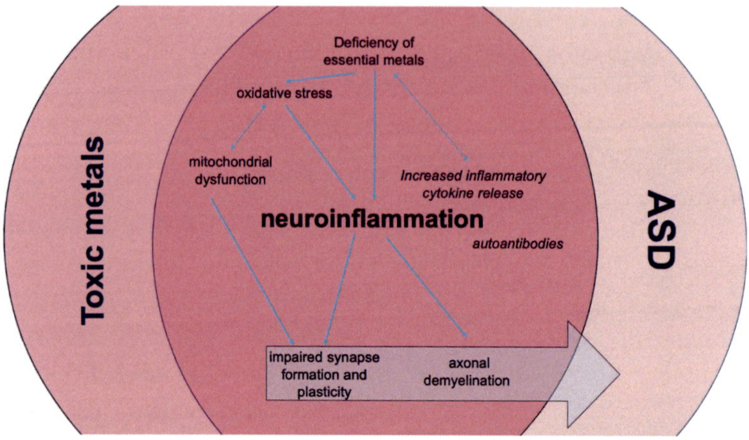

Figure 5.2 ***Shared pathologies of exposure to toxic trace metals and ASD:*** ASDs are characterized by impaired synapse formation and plasticity during brain development, as well as white matter (axonal) abnormalities that, together, may lead to altered connectivity of the developing CNS. As underlying causes, among others, inflammatory and immune system activating processes, mitochondrial dysfunction, and oxidative stress have been discussed. The latter can be triggered by direct effects of toxic metals and indirectly due to deficiencies of essentials metals competing with toxic metals, which reveals a significant overlap between the effects of toxic metals and pathologies observed in ASD. *ASD*, autism spectrum disorder.

those recurring motifs is inflammation/immune activation, which can be caused by toxic metals through different mechanisms such as altered gene expression and competition with essential metals, mitochondrial dysfunction, and increased oxidative stress. Inflammation/immune activation will ultimately contribute to impaired axonal myelination, synapse formation, and function (Fig. 5.2).

References

Amitai, Y., Katz, D., Lifshitz, M., Gofin, R., Tepferberg, M., Almog, S., 1999. Prenatal lead exposure in Israel: an international comparison. Isr Med Assoc J 1 (4), 250–253.

Aschner, M., Aschner, J.L., 1990. Mercury neurotoxicity: mechanisms of blood-brain barrier transport. Neurosci Biobehav Rev 14 (2), 169–176.

Aschner, M., Yao, C.P., Allen, J.W., Tan, K.H., 2000. Methylmercury alters glutamate transport in astrocytes. Neurochem Int 37 (2–3), 199–206.

Bakhti, M., Aggarwal, S., Simons, M., 2014. Myelin architecture: zippering membranes tightly together. Cell Mol Life Sci 71 (7), 1265–1277.

Ballatori, N., Clarkson, T.W., 1984. Dependence of biliary secretion of inorganic mercury on the biliary transport of glutathione. Biochem Pharmacol 33 (7), 1093–1098.

Bartosiewicz, M.J., Jenkins, D., Penn, S., Emery, J., Buckpitt, A., 2001. Unique gene expression patterns in liver and kidney associated with exposure to chemical toxicants. J Pharmacol Exp Therapeut 297 (3), 895–905.

Bernal, J., 2007. Thyroid hormone receptors in brain development and function. Nat Clin Pract Endocrinol Metabol 3 (3), 249–259.
Bertin, G., Averbeck, D., 2006. Cadmium: cellular effects, modifications of biomolecules, modulation of DNA repair and genotoxic consequences (a review). Biochimie 88 (11), 1549–1559.
Binder, D.K., Scharfman, H.E., 2004. Brain-derived neurotrophic factor. Growth Factors 22, 123–131.
Birben, E., Sahiner, U.M., Sackesen, C., Erzurum, S., Kalayci, O., 2012. Oxidative stress and antioxidant defense. World Allergy Organ J 5 (1), 9–19.
Bjørklund, G., Meguid, N.A., El-Bana, M.A., Tinkov, A.A., Saad, K., Dadar, M., Hemimi, M., Skalny, A.V., Hosnedlová, B., Kizek, R., Osredkar, J., Urbina, M.A., Fabjan, T., El-Houfey, A.A., Kałużna-Czaplińska, J., Gątarek, P., Chirumbolo, S., 2020a. Oxidative stress in autism spectrum disorder. Mol Neurobiol 57 (5), 2314–2332.
Bjørklund, G., Peana, M., Dadar, M., Chirumbolo, S., Aaseth, J., Martins, N., 2020b. Mercury-induced autoimmunity: drifting from micro to macro concerns on autoimmune disorders. Clin Immunol 108352.
Bórquez, D.A., Urrutia, P.J., Wilson, C., van Zundert, B., Núñez, M.T., González-Billault, C., 2016. Dissecting the role of redox signaling in neuronal development. J Neurochem 137 (4), 506–517.
Bressler, J.P., Goldstein, G.W., 1991. Mechanisms of lead neurotoxicity. Biochem Pharmacol 41 (4), 479–484.
Bressler, J., Kim, K.A., Chakraborti, T., Goldstein, G., 1999. Molecular mechanisms of lead neurotoxicity. Neurochem Res 24 (4), 595–600.
Breton, J., Le Clère, K., Daniel, C., Sauty, M., Nakab, L., Chassat, T., Dewulf, J., Penet, S., Carnoy, C., Thomas, P., Pot, B., Nesslany, F., Foligné, B., 2013. Chronic ingestion of cadmium and lead alters the bioavailability of essential and heavy metals, gene expression pathways and genotoxicity in mouse intestine. Arch Toxicol 87 (10), 1787–1795.
Cariccio, V.L., Samà, A., Bramanti, P., Mazzon, E., 2019. Mercury involvement in neuronal damage and in neurodegenerative diseases. Biol Trace Elem Res 187 (2), 341–356.
Carocci, A., Rovito, N., Sinicropi, M.S., Genchi, G., 2014. Mercury toxicity and neurodegenerative effects. Rev Environ Contam Toxicol 229, 1–18.
Carvalho, C.M., Chew, E.H., Hashemy, S.I., Lu, J., Holmgren, A., 2008. Inhibition of the human thioredoxin system. A molecular mechanism of mercury toxicity. J Biol Chem 283 (18), 11913–11923.
Castora, F.J., 2019. Mitochondrial function and abnormalities implicated in the pathogenesis of ASD. Prog Neuro-Psychopharmacol Biol Psychiatry 92, 83–108.
Cecil, K.M., Brubaker, C.J., Adler, C.M., Dietrich, K.N., Altaye, M., Egelhoff, J.C., Wessel, S., Elangovan, I., Hornung, R., Jarvis, K., Lanphear, B.P., 2008. Decreased brain volume in adults with childhood lead exposure. PLoS Med 5 (5), e112.
Chauhan, A., Audhya, T., Chauhan, V., 2012. Brain region-specific glutathione redox imbalance in autism. Neurochem Res 37 (8), 1681–1689.
Chen, Z., Myers, R., Wei, T., Bind, E., Kassim, P., Wang, G., Ji, Y., Hong, X., Caruso, D., Bartell, T., Gong, Y., Strickland, P., Navas-Acien, A., Guallar, E., Wang, X., 2014. Placental transfer and concentrations of cadmium, mercury, lead, and selenium in mothers, newborns, and young children. J Expo Sci Environ Epidemiol 24 (5), 537–544.
Chibowska, K., Baranowska-Bosiacka, I., Falkowska, A., Gutowska, I., Goschorska, M., Chlubek, D., 2016. Effect of lead (Pb) on inflammatory processes in the brain. Int J Mol Sci 17 (12)
Coria, F., Berciano, M.T., Berciano, J., Lafarga, M., 1984. Axon membrane remodeling in the lead-induced demyelinating neuropathy of the rat. Brain Res 291 (2), 369–372.

Dabrowska-Bouta, B., Sulkowski, G., Walski, M., Struzyńska, L., Lenkiewicz, A., Rafałowska, U., 2000. Acute lead intoxication in vivo affects myelin membrane morphology and CNPase activity. Exp Toxicol Pathol 52 (3), 257–263.

Deth, R., Muratore, C., Benzecry, J., Power-Charnitsky, V.A., Waly, M., 2008. How environmental and genetic factors combine to cause autism: a redox/methylation hypothesis. Neurotoxicology 29 (1), 190–201.

Devine, M.J., Kittler, J.T., 2018. Mitochondria at the neuronal presynapse in health and disease. Nat Rev Neurosci 19 (2), 63–80.

Dhillon, S., Hellings, J.A., Butler, M.G., 2011. Genetics and mitochondrial abnormalities in autism spectrum disorders: a review. Curr Genom 12 (5), 322–332.

Dong, S., Shen, H.M., Ong, C.N., 2001. Cadmium-induced apoptosis and phenotypic changes in mouse thymocytes. Mol Cell Biochem 222 (1–2), 11–20.

Dietert, R.R., Dietert, J.M., Dewitt, J.C., 2011. Environmental risk factors for autism. Emerg Health Threats J 4, 7111.

Edmiston, E., Ashwood, P., Van de Water, J., 2017. Autoimmunity, autoantibodies, and autism spectrum disorder. Biol Psychiatr 81 (5), 383–390.

El-Fawal, H.A., Waterman, S.J., De Feo, A., Shamy, M.Y., 1999. Neuroimmunotoxicology: humoral assessment of neurotoxicity and autoimmune mechanisms. Environ Health Perspect 107 (Suppl. 5), 767–775.

Eyssen-Hernandez, R., Ladoux, A., Frelin, C., 1996. Differential regulation of cardiac heme oxygenase-1 and vascular endothelial growth factor mRNA expressions by hemin, heavy metals, heat shock and anoxia. FEBS Lett 382 (3), 229–233.

Falluel-Morel, A., Sokolowski, K., Sisti, H.M., Zhou, X., Shors, T.J., Dicicco-Bloom, E., 2007. Developmental mercury exposure elicits acute hippocampal cell death, reductions in neurogenesis, and severe learning deficits during puberty. J Neurochem 103 (5), 1968–1981.

Farina, M., Aschner, M., 2019. Glutathione antioxidant system and methylmercury-induced neurotoxicity: an intriguing interplay. Biochim Biophys Acta Gen Subj 1863 (12), 129285.

Farina, M., Frizzo, M.E., Soares, F.A., Schwalm, F.D., Dietrich, M.O., Zeni, G., Rocha, J.B., Souza, D.O., 2003. Ebselen protects against methylmercury-induced inhibition of glutamate uptake by cortical slices from adult mice. Toxicol Lett 144 (3), 351–357.

Farina, M., Rocha, J.B., Aschner, M., 2011. Mechanisms of methylmercury-induced neurotoxicity: evidence from experimental studies. Life Sci 89 (15–16), 555–563.

Flora, S.J., Saxena, G., Gautam, P., Kaur, P., Gill, K.D., 2007. Response of lead-induced oxidative stress and alterations in biogenic amines in different rat brain regions to combined administration of DMSA and MiADMSA. Chem Biol Interact 170 (3), 209–220.

Flora, S.J., Singh, S., Tandon, S.K., 1989. Thiamine and zinc in prevention or therapy of lead intoxication. J Int Med Res 17 (1), 68–75.

Flora, S.J., Kumar, D., Das Gupta, S., 1991. Interaction of zinc, methionine or their combination with lead at gastrointestinal or post-absorptive level in rats. Pharmacol Toxicol 68 (1), 3–7.

Figueiredo-Pereira, M.E., Yakushin, S., Cohen, G., 1998. Disruption of the intracellular sulfhydryl homeostasis by cadmium-induced oxidative stress leads to protein thiolation and ubiquitination in neuronal cells. J Biol Chem 273 (21), 12703–12709.

Finsterer, J., Scorza, F.A., 2017. Renal manifestations of primary mitochondrial disorders. Biomed Rep 6 (5), 487–494.

Forder, J.P., Tymianski, M., 2009. Postsynaptic mechanisms of excitotoxicity: involvement of postsynaptic density proteins, radicals, and oxidant molecules. Neuroscience 158 (1), 293–300.

Franciscato, C., Moraes-Silva, L., Duarte, F.A., Oliveira, C.S., Ineu, R.P., Flores, E.M., Dressler, V.L., Peixoto, N.C., Pereira, M.E., 2011. Delayed biochemical changes induced by mercury intoxication are prevented by zinc pre-exposure. Ecotoxicol Environ Saf 74 (3), 480–486.

Franco, J.L., Posser, T., Dunkley, P.R., Dickson, P.W., Mattos, J.J., Martins, R., Bainy, A.C., Marques, M.R., Dafre, A.L., Farina, M., 2009. Methylmercury neurotoxicity is associated with inhibition of the antioxidant enzyme glutathione peroxidase. Free Radic Biol Med 47 (4), 449–457.

Frye, R.E., Rossignol, D.A., 2011. Mitochondrial dysfunction can connect the diverse medical symptoms associated with autism spectrum disorders. Pediatr Res 69 (5 Pt 2), 41R–7R.

Geier, D.A., Audhya, T., Kern, J.K., Geier, M.R., 2010. Blood mercury levels in autism spectrum disorder: is there a threshold level? Acta Neurobiol Exp 70 (2), 177–186.

Gemma, C., Vila, J., Bachstetter, A., et al., 2007. Oxidative stress and the aging brain: from theory to prevention. In: Riddle, D.R. (Ed.), Brain Aging: Models, Methods, and Mechanisms. CRC Press/Taylor & Francis, Boca Raton (FL) (Chapter 15).

Ghanizadeh, A., 2012. Malondialdehyde, Bcl-2, superoxide dismutase and glutathione peroxidase may mediate the association of sonic hedgehog protein and oxidative stress in autism. Neurochem Res 37 (4), 899–901.

Ghanizadeh, A., Akhondzadeh, S., Hormozi, M., Makarem, A., Abotorabi-Zarchi, M., Firoozabadi, A., 2012. Glutathione-related factors and oxidative stress in autism, a review. Curr Med Chem 19 (23), 4000–4005.

Giulivi, C., Zhang, Y.F., Omanska-Klusek, A., Ross-Inta, C., Wong, S., Hertz-Picciotto, I., Tassone, F., Pessah, I.N., 2010. Mitochondrial dysfunction in autism. J Am Med Assoc 304 (21), 2389–2396.

Griffiths, K.K., Levy, R.J., 2017. Evidence of mitochondrial dysfunction in autism: biochemical links, genetic-based associations, and non-energy-related mechanisms. Oxid Med Cell Longev 2017, 4314025.

Gu, Y., Dee, C.M., Shen, J., 2011. Interaction of free radicals, matrix metalloproteinases and caveolin-1 impacts blood-brain barrier permeability. Front Biosci 3, 1216–1231.

Guzzi, G., La Porta, C.A., 2008. Molecular mechanisms triggered by mercury. Toxicology 244 (1), 1–12.

Hughes, H.K., Mills Ko, E., Rose, D., Ashwood, P., 2018. Immune dysfunction and autoimmunity as pathological mechanisms in autism spectrum disorders. Front Cell Neurosci 12, 405.

Ishitobi, H., Mori, K., Yoshida, K., Watanabe, C., 2007. Effects of perinatal exposure to low-dose cadmium on thyroid hormone-related and sex hormone receptor gene expressions in brain of offspring. Neurotoxicology 28 (4), 790–797.

Jacobo-Estrada, T., Santoyo-Sánchez, M., Thévenod, F., Barbier, O., 2017. Cadmium handling, toxicity and molecular targets involved during pregnancy: lessons from experimental models. Int J Mol Sci 18 (7).

James, S.J., Rose, S., Melnyk, S., Jernigan, S., Blossom, S., Pavliv, O., Gaylor, D.W., 2009. Cellular and mitochondrial glutathione redox imbalance in lymphoblastoid cells derived from children with autism. FASEB J 23 (8), 2374–2383.

James, S.J., Cutler, P., Melnyk, S., Jernigan, S., Janak, L., Gaylor, D.W., Neubrander, J.A., 2004. Metabolic biomarkers of increased oxidative stress and impaired methylation capacity in children with autism. Am J Clin Nutr 80 (6), 1611–1617.

Kany, S., Vollrath, J.T., Relja, B., 2019. Cytokines in inflammatory disease. Int J Mol Sci 20 (23).

Kasahara, E., Inoue, M., 2015. Cross-talk between HPA-axis-increased glucocorticoids and mitochondrial stress determines immune responses and clinical manifestations of patients with sepsis. Redox Rep 20 (1), 1–10.

Kern, J.K., Geier, D.A., Audhya, T., King, P.G., Sykes, L.K., Geier, M.R., 2012. Evidence of parallels between mercury intoxication and the brain pathology in autism. Acta Neurobiol Exp 72 (2), 113–153.
Kern, J.K., Geier, D.A., Homme, K.G., Geier, M.R., 2020. Examining the evidence that ethylmercury crosses the blood-brain barrier. Environ Toxicol Pharmacol 74, 103312.
Kern, J.K., Jones, A.M., 2006. Evidence of toxicity, oxidative stress, and neuronal insult in autism. J Toxicol Environ Health B Crit Rev 9 (6), 485–499.
Kerper, L.E., Hinkle, P.M., 1997. Lead uptake in brain capillary endothelial cells: activation by calcium store depletion. Toxicol Appl Pharmacol 146 (1), 127–133.
Kippler, M., Hoque, A.M., Raqib, R., Ohrvik, H., Ekström, E.C., Vahter, M., 2010. Accumulation of cadmium in human placenta interacts with the transport of micronutrients to the fetus. Toxicol Lett 192 (2), 162–168.
Knapp, L.T., Klann, E., 2002. Role of reactive oxygen species in hippocampal long-term potentiation: contributory or inhibitory? J Neurosci Res 70 (1), 1–7.
Lasley, S.M., Green, M.C., Gilbert, M.E., 1999. Influence of exposure period on in vivo hippocampal glutamate and GABA release in rats chronically exposed to lead. Neurotoxicology 20 (4), 619–629.
Lasley, S.M., Green, M.C., Gilbert, M.E., 2001. Rat hippocampal NMDA receptor binding as a function of chronic lead exposure level. Neurotoxicol Teratol 23 (2), 185–189.
Lazar, M., Miles, L.M., Babb, J.S., Donaldson, J.B., 2014. Axonal deficits in young adults with high functioning autism and their impact on processing speed. Neuroimage Clin 4, 417–425.
Lidsky, T.I., Schneider, J.S., 2003. Lead neurotoxicity in children: basic mechanisms and clinical correlates. Brain 126 (Pt 1), 5–19.
Lidsky, T.I., Schneider, J.S., 2006. Adverse effects of childhood lead poisoning: the clinical neuropsychological perspective. Environ Res 100 (2), 284–293.
Lim, J.L., Wilhelmus, M.M., de Vries, H.E., Drukarch, B., Hoozemans, J.J., van Horssen, J., 2014. Antioxidative defense mechanisms controlled by Nrf2: state-of-the-art and clinical perspectives in neurodegenerative diseases. Arch Toxicol 88 (10), 1773–1786.
Limongelli, G., Masarone, D., Pacileo, G., 2017. Mitochondrial disease and the heart. Heart 103 (5), 390–398.
Lindahl, L.S., Bird, L., Legare, M.E., Mikeska, G., Bratton, G.R., Tiffany-Castiglioni, E., 1999. Differential ability of astroglia and neuronal cells to accumulate lead: dependence on cell type and on degree of differentiation. Toxicol Sci 50 (2), 236–243.
Lingappan, K., 2018. NF-κB in oxidative stress. Curr Opin Toxicol 7, 81–86.
Liu, H., Oteiza, P.I., Gershwin, M.E., Golub, M.S., Keen, C.L., 1992. Effects of maternal marginal zinc deficiency on myelin protein profiles in the suckling rat and infant rhesus monkey. Biol Trace Elem Res 34 (1), 55–66.
Malik, M., Tauqeer, Z., Sheikh, A.M., Wen, G., Nagori, A., Yang, K., Brown, W.T., Li, X., 2011. NF-κB signaling in the brain of autistic subjects. Mediat Inflamm 2011, 785265.
Maiuolo, J., Macrì, R., Bava, I., Gliozzi, M., Musolino, V., Nucera, S., Carresi, C., Scicchitano, M., Bosco, F., Scarano, F., Palma, E., Gratteri, S., Mollace, V., 2019. Myelin disturbances produced by sub-toxic concentration of heavy metals: the role of oligodendrocyte dysfunction. Int J Mol Sci 20 (18).
Manivasagam, T., Arunadevi, S., Essa, M.M., SaravanaBabu, C., Borah, A., Thenmozhi, A.J., Qoronfleh, M.W., 2020. Role of oxidative stress and antioxidants in autism. Adv Neurobiol 24, 193–206.
McCoy, L., Richfield, E.K., Cory-Slechta, D.A., 1997. Regional decreases in alpha-[3H]amino-3-hydroxy-5-methylisoxazole-4-propionic acid ([3H]AMPA) and 6-[3H]cyano-7-nitroquinoxaline-2,3-dione ([3H]CNQX) binding in response to chronic low-level

lead exposure: reversal versus potentiation by chronic dopamine agonist treatment. J Neurochem 69 (6), 2466–2476.

Mishra, K.P., 2009. Lead exposure and its impact on immune system: a review. Toxicol Vitro 23 (6), 969–972.

Monnet-Tschudi, F., Zurich, M.G., Boschat, C., Corbaz, A., Honegger, P., 2006. Involvement of environmental mercury and lead in the etiology of neurodegenerative diseases. Rev Environ Health 21 (2), 105–117.

Moretto, M.B., Funchal, C., Santos, A.Q., Gottfried, C., Boff, B., Zeni, G., Pureur, R.P., Souza, D.O., Wofchuk, S., Rocha, J.B., 2005. Ebselen protects glutamate uptake inhibition caused by methyl mercury but does not by Hg^{2+}. Toxicology 214 (1–2), 57–66.

Morris, G., Puri, B.K., Frye, R.E., Maes, M., 2018. The putative role of environmental mercury in the pathogenesis and pathophysiology of autism spectrum disorders and subtypes. Mol Neurobiol 55 (6), 4834–4856.

Motts, J.A., Shirley, D.L., Silbergeld, E.K., Nyland, J.F., 2014. Novel biomarkers of mercury-induced autoimmune dysfunction: a cross-sectional study in Amazonian Brazil. Environ Res 132, 12–18.

Moulis, J.M., 2010. Cellular mechanisms of cadmium toxicity related to the homeostasis of essential metals. Biometals 23 (5), 877–896.

Napolioni, V., Persico, A.M., Porcelli, V., Palmieri, L., 2011. The mitochondrial aspartate/glutamate carrier AGC1 and calcium homeostasis: physiological links and abnormalities in autism. Mol Neurobiol 44 (1), 83–92.

Neal, A.P., Stansfield, K.H., Worley, P.F., Thompson, R.E., Guilarte, T.R., 2010. Lead exposure during synaptogenesis alters vesicular proteins and impairs vesicular release: potential role of NMDA receptor-dependent BDNF signaling. Toxicol Sci 116 (1), 249–263.

Nguyen, R.L., Medvedeva, Y.V., Ayyagari, T.E., Schmunk, G., Gargus, J.J., 2018. Intracellular calcium dysregulation in autism spectrum disorder: an analysis of converging organelle signaling pathways. Biochim Biophys Acta Mol Cell Res 1865 (11 Pt B), 1718–1732.

Niciu, M.J., Kelmendi, B., Sanacora, G., 2012. Overview of glutamatergic neurotransmission in the nervous system. Pharmacol Biochem Behav 100 (4), 656–664.

Pamphlett, R., Kum Jew, S., 2019. Mercury is taken up selectively by cells involved in joint, bone, and connective tissue disorders. Front Med 6, 168.

Pieper, I., Wehe, C.A., Bornhorst, J., Ebert, F., Leffers, L., Holtkamp, M., Höseler, P., Weber, T., Mangerich, A., Bürkle, A., Karst, U., Schwerdtle, T., 2014. Mechanisms of Hg species induced toxicity in cultured human astrocytes: genotoxicity and DNA-damage response. Metallomics 6 (3), 662–671.

Pollard, K.M., Cauvi, D.M., Toomey, C.B., Hultman, P., Kono, D.H., 2019. Mercury-induced inflammation and autoimmunity. Biochim Biophys Acta Gen Subj 1863 (12), 129299.

Porciúncula, L.O., Rocha, J.B., Tavares, R.G., Ghisleni, G., Reis, M., Souza, D.O., 2003. Methylmercury inhibits glutamate uptake by synaptic vesicles from rat brain. Neuroreport 14 (4), 577–580.

Pun, P.B., Lu, J., Moochhala, S., 2009. Involvement of ROS in BBB dysfunction. Free Radic Res 43 (4), 348–364.

Ramana, K.V., Srivastava, S., Singhal, S.S., 2013. Lipid peroxidation products in human health and disease. Oxid Med Cell Longev 2013, 583438.

Reynolds, J.N., Racz, W.J., 1987. Effects of methylmercury on the spontaneous and potassium-evoked release of endogenous amino acids from mouse cerebellar slices. Can J Physiol Pharmacol 65 (5), 791–798.

Rodier, P.M., 1995. Developing brain as a target of toxicity. Environ Health Perspect 103 (Suppl. 6), 73–76.
Rossignol, D.A., Frye, R.E., 2012. Mitochondrial dysfunction in autism spectrum disorders: a systematic review and meta-analysis. Mol Psychiatr 17 (3), 290–314.
Rubenstein, J.L., Merzenich, M.M., 2003. Model of autism: increased ratio of excitation/inhibition in key neural systems. Gene Brain Behav 2 (5), 255–267.
Sakamoto, M., Tatsuta, N., Izumo, K., Phan, P.T., Vu, L.D., Yamamoto, M., Nakamura, M., Nakai, K., Murata, K., 2018. Health impacts and biomarkers of prenatal exposure to methylmercury: lessons from Minamata, Japan. Toxics 6 (3).
Salim, S., 2017. Oxidative stress and the central nervous system. J Pharmacol Exp Therapeut 360 (1), 201–205.
Segerstråle, M., Juuri, J., Lanore, F., Piepponen, P., Lauri, S.E., Mulle, C., Taira, T., 2010. High firing rate of neonatal hippocampal interneurons is caused by attenuation of afterhyperpolarizing potassium currents by tonically active kainate receptors. J Neurosci 30 (19), 6507–6514.
Saghazadeh, A., Rezaei, N., 2017. Brain-derived neurotrophic factor levels in autism: a systematic review and meta-analysis. J Autism Dev Disord 47 (4), 1018–1029.
Shanker, G., Syversen, T., Aschner, M., 2003. Astrocyte-mediated methylmercury neurotoxicity. Biol Trace Elem Res 95 (1), 1–10.
Shen, H.Y., Huang, N., Reemmer, J., Xiao, L., 2018. Adenosine actions on oligodendroglia and myelination in autism spectrum disorder. Front Cell Neurosci 12, 482.
Shukla, G.S., Hussain, T., Chandra, S.V., 1987. Possible role of regional superoxide dismutase activity and lipid peroxide levels in cadmium neurotoxicity: in vivo and in vitro studies in growing rats. Life Sci 41 (19), 2215–2221.
Siniscalco, D., Schultz, S., Brigida, A.L., Antonucci, N., 2018. Inflammation and neuro-immune dysregulations in autism spectrum disorders. Pharmaceuticals 11 (2).
Takahashi, T., Shimohata, T., 2019. Vascular dysfunction induced by mercury exposure. Int J Mol Sci 20 (10).
Theoharides, T.C., Tsilioni, I., Patel, A.B., Doyle, R., 2016. Atopic diseases and inflammation of the brain in the pathogenesis of autism spectrum disorders. Transl Psychiatry 6 (6), e844.
Thévenod, F., 2009. Cadmium and cellular signaling cascades: to be or not to be? Toxicol Appl Pharmacol 238 (3), 221–239.
Thomason, M.E., Hect, J.L., Rauh, V.A., Trentacosta, C., Wheelock, M.D., Eggebrecht, A.T., Espinoza-Heredia, C., Burt, S.A., 2019. Prenatal lead exposure impacts cross-hemispheric and long-range connectivity in the human fetal brain. Neuroimage 191, 186–192.
Tiffany-Castiglion, E., Qian, Y., 2001. Astroglia as metal depots: molecular mechanisms for metal accumulation, storage and release. Neurotoxicology 22 (5), 577–592.
Toscano, C.D., Guilarte, T.R., 2005. Lead neurotoxicity: from exposure to molecular effects. Brain Res Brain Res Rev 49 (3), 529–554.
Tsang, D., Tsang, Y.S., Ho, W.K., Wong, R.N., 1997. Myelin basic protein is a zinc-binding protein in brain: possible role in myelin compaction. Neurochem Res 22 (7), 811–819.
Valiente-Pallejà, A., Torrell, H., Muntané, G., Cortés, M.J., Martínez-Leal, R., Abasolo, N., Alonso, Y., Vilella, E., Martorell, L., 2018. Genetic and clinical evidence of mitochondrial dysfunction in autism spectrum disorder and intellectual disability. Hum Mol Genet 27 (5), 891–900.
Varga, N.Á., Pentelényi, K., Balicza, P., Gézsi, A., Reményi, V., Hársfalvi, V., Bencsik, R., Illés, A., Prekop, C., Molnár, M.J., 2018. Mitochondrial dysfunction and autism: comprehensive genetic analyses of children with autism and mtDNA deletion. Behav Brain Funct 14 (1), 4.

Vargas, D.L., Nascimbene, C., Krishnan, C., Zimmerman, A.W., Pardo, C.A., 2005. Neuroglial activation and neuroinflammation in the brain of patients with autism. Ann Neurol 57 (1), 67–81.
Vas, J., Monestier, M., 2008. Immunology of mercury. Ann N Y Acad Sci 1143, 240–267.
Vendrell, I., Carrascal, M., Vilaro, M.T., Abian, J., Rodriguez-Farre, E., Sunol, C., 2007. Cell viability and proteomic analysis in cultured neurons exposed to methylmercury. Hum Exp Toxicol 26 (4), 263–272.
Villeda-Hernández, J., Barroso-Moguel, R., Méndez-Armenta, M., Nava-Ruíz, C., Huerta-Romero, R., Ríos, C., 2001. Enhanced brain regional lipid peroxidation in developing rats exposed to low level lead acetate. Brain Res Bull 55 (2), 247–251.
Wei, H., Alberts, I., Li, X., 2013. Brain IL-6 and autism. Neuroscience 252, 320–325.
Wei, H., Chadman, K.K., McCloskey, D.P., Sheikh, A.M., Malik, M., Brown, W.T., Li, X., 2012b. Brain IL-6 elevation causes neuronal circuitry imbalances and mediates autism-like behaviors. Biochim Biophys Acta 1822 (6), 831–842.
Wei, H., Mori, S., Hua, K., Li, X., 2012a. Alteration of brain volume in IL-6 overexpressing mice related to autism. Int J Dev Neurosci 30 (7), 554–559.
Weissman, J.R., Kelley, R.I., Bauman, M.L., Cohen, B.H., Murray, K.F., Mitchell, R.L., Kern, R.L., Natowicz, M.R., 2008. Mitochondrial disease in autism spectrum disorder patients: a cohort analysis. PLoS One 3 (11), e3815.
Yang, P.M., Chiu, S.J., Lin, K.A., Lin, L.Y., 2004. Effect of cadmium on cell cycle progression in Chinese hamster ovary cells. Chem Biol Interact 149 (2–3), 125–136.
Yang, H., Shu, Y., 2015. Cadmium transporters in the kidney and cadmium-induced nephrotoxicity. Int J Mol Sci 16 (1), 1484–1494.
Yin, Z., Milatovic, D., Aschner, J.L., Syversen, T., Rocha, J.B., Souza, D.O., Sidoryk, M., Albrecht, J., Aschner, M., 2007. Methylmercury induces oxidative injury, alterations in permeability and glutamine transport in cultured astrocytes. Brain Res 1131 (1), 1–10.
Zhai, Q., Cen, S., Jiang, J., Zhao, J., Zhang, H., Chen, W., 2019. Disturbance of trace element and gut microbiota profiles as indicators of autism spectrum disorder: a pilot study of Chinese children. Environ Res 171, 501–509.

CHAPTER 6

Biometals and nutrition in autism spectrum disorders

Contents

Absorption and transport of metals

Essential trace metals are absorbed in the gastrointestinal (GI) system. As charged elements, metal ions cannot passively diffuse across lipid membranes. Therefore, dedicated transport systems control the uptake into enterocytes in the GI tract and the release of ions from these cells into the blood circulation.

Iron

Iron absorption takes place in the duodenum (Garrick, 2011). The transporter that mediates the uptake into enterocytes on the apical side is the divalent metal transporter (DMT1). Therefore, iron must be maintained or reduced to its divalent (Fe^{2+}) form by the ferrireductase duodenal cytochrome b (Dcytb). Once taken up, iron can be stored by the iron storage protein ferritin or shuttled through the cell. Iron is released on the basal side by the transporter ferroportin-1 (Fpn; SLC40A1). In addition, iron oxidation from Fe^{2+} to Fe^{3+} by hephaestin (Hp) or ceruloplasmin (Cp) must occur, for iron to be able to bind to transferrin (Tf) (Garrick and Garrick, 2009; McCarthy and Kosman, 2015) (Fig. 6.1).

Biometals in Autism Spectrum Disorders
ISBN 978-0-12-821132-8
https://doi.org/10.1016/B978-0-12-821132-8.00006-9

Once in the blood circulation, iron is transported bound to Tf and taken up into cells that express the Tf receptor (TfR) on their surface. Through this process, iron is moved to the liver that can act as an iron store and supplies iron to other tissues such as the bone marrow, where large quantities are needed to form hemoglobin, but also the brain. Iron crosses the blood–brain barrier (BBB) by uptake and intracellular trafficking of the iron/transferrin (Tf) through interaction with the TfR (McCarthy and Kosman, 2015). Once endocytosed by cells of the BBB, the endosomal iron/Tf complex will release iron, and iron will subsequently exit the endosome through DMT1. Inside the cytoplasm, divalent iron can bind to iron-buffering proteins such as poly(RC)-binding protein 2 (PCBP2) or ferritin. On the basal side of the BBB cell, iron gets released by Fpn.

Iron is provided to the developing embryo through similar transport mechanisms using DMT1 and TfR that are present at the placenta (Cetin et al., 2011). Tf/iron is thought to be the predominant iron source taken up by the placenta, and after endosomal release, Fpn mediates its export (Sangkhae and Nemeth, 2019).

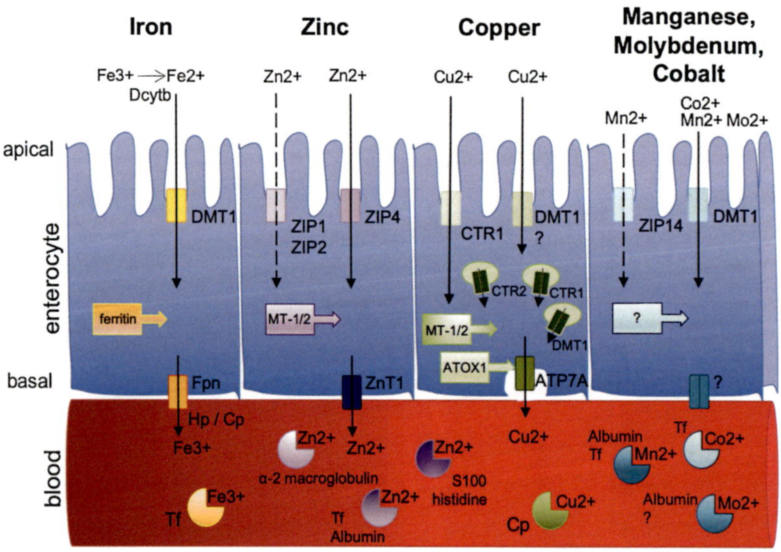

Figure 6.1 ***GI absorption of essential trace metals.*** Trace metals are taken up at the apical side from dietary sources in the lumen of the GI tract. Inside enterocytes, specific metallochaperones mediate the transport of metal ions through the cytoplasm. At the basal side, metals are released into the blood circulation by active transport mechanisms and transported in serum bound to proteins. However, serum proteins such as Tf and albumin can bind several different trace metals. Similarly, DMT1 can import different divalent metals. GI, gastrointestinal.

Zinc

Zinc is taken up into enterocytes in the proximal small intestine, most likely the jejunum. The human genome encodes for at least 24 zinc transport proteins that are found in two families. Ten zinc transporters of the ZnT (SLC30A) family mediate the release of zinc from the cytoplasm out of the cell or mediate sequestration of zinc into intracellular organelles, and 14 Zrt- and Irt-like protein transporters of the ZIP (SLC39A) family import zinc into the cytoplasm from the extracellular space or intracellular organelles. However, not all transporters are expressed in the GI tract. Both the ZnT and ZIP transporter families exhibit unique tissue-specific expression (Roohani et al., 2013). In the GI tract, the ZIP4 (SLC39A4) protein, in particular, is critical for uptake in the small intestine (Cousins and McMahon, 2000). ZIP4 is localized at the apical membrane of enterocytes in the entire GI tract and imports zinc into the cytoplasm. Besides, ZIP1 (SLC39A1) and ZIP2 (SLC39A2) have been reported to contribute to zinc uptake into enterocytes.

"Free" zinc is found in very low concentration inside cells (in the picomolar range). In contrast, total zinc levels are around 200–300 μM (Maret, 2015). Thus, the majority of intracellular zinc is protein bound. In enterocytes, as in many other cell types, metallothioneins (MTs) act as zinc buffering and transport proteins. Four major MT isoforms (MT-1–MT-4) have been identified in humans. While MT-1 and MT-2 are widely expressed, including in the GI tract, MT-3 is a brain-specific protein. MT-4 expression in humans is currently not well known, but the protein may be present in squamous epithelia (Coyle et al., 2002; Li and Maret, 2008; Levaot and Hershfinkel, 2018). Although MTs play a significant role in zinc transport through the enterocyte, many zinc-binding proteins exist in cells that may additionally contribute to this process. On the basal side of the enterocyte, zinc is released into the blood circulation mainly through the zinc transporter ZnT1 (SLC30A1) (Cousins, 2010) (Fig. 6.1). There, zinc binds to proteins such as albumin, α-2 macroglobulin, histidine, or proteins of the S100 family, and transferrin (Chesters and Will, 1981) for its distribution through the blood circulation. It is mainly transported by albumin (Scott and Bradwell, 1983).

Zinc transport across other barriers is mediated by ZnTs and ZIPs as well. For example, ZnT1, ZnT2, ZnT4, and ZnT5 were discussed in relation to zinc transport across the placenta, while DMT1 (divalent metal transporter 1/SLC11A2) may import zinc into brain capillary endothelial cells where other unidentified zinc transporters mediate the export for BBB crossing (Takeda, 2000).

Copper

Copper is absorbed mainly in the stomach and duodenum (van den Berghe and Klomp., 2009). The sequence of events by which dietary copper is absorbed is currently not completely understood. However, two transport proteins, the high-affinity copper transporter 1 (CTR1; SLC31A1) on the apical side and ATP7A on the basal side, seem to be critical for copper absorption.

On the apical side, DMT1 and the low-affinity copper transporter 2 (CTR2; SLC31A2) may also aid in copper uptake into the enterocyte (Arredondo et al., 2003; Bertinato et al., 2008). CTR1, CTR2, and DMT1 may be endocytosed in a copper concentration-dependent way. Copper is utilized and transported preferably in its divalent (Cu^{2+}) state in the human body, since monocationic copper (Cu^{+}) is highly reactive and may increase oxidative stress significantly. The concentrations of free Cu^{+} and Cu^{2+} are kept to a minimum (in the attomolar (10^{-18} M) range) (Rae et al., 1999) by high-affinity copper-binding proteins (copper chaperones) (Robinson and Winge, 2010). Thus, once inside the cell, copper is bound by cytosolic proteins such as MTs and ATOX1 (Hamza et al., 2003) that act as a buffer. ATOX1 regulates the intracellular transport of copper to ATP7A, which imports copper into the trans-Golgi network, where it binds to further proteins that are ultimately secreted. In addition, ATP7A can translocate to the basal membrane and release copper into the blood circulation (van den Berghe and Klomp., 2009).

Once in the blood, copper is mainly bound by ceruloplasmin (Cp) that, however, may not participate in a copper exchange between serum and tissues, but also a2-macroglobulin, albumin, or the amino acid histidine (van den Berghe and Klomp., 2009) (Fig. 6.1). Interestingly, Cp levels were reported altered in ASD in several studies (Goldani et al., 2014). Similar to iron, newly absorbed copper is transported to the liver and stored in hepatocytes. From there, it can be released and redistributed throughout the body at a later time. Besides, Cp is degraded in the liver, and the copper released by the process is available for redistribution. Thus, indirectly, Cp is involved in copper transport.

Similar mechanisms may also regulate copper translocation into other tissues. For example, ATP7A mRNA expression is highest in placental tissue (Paynter et al., 1994), where ATP7A, together with CTR1 and DMT1, may have similar functions as in enterocytes (McArdle et al., 2008). However, other copper transport proteins such as ATP7B can be expressed in addition at barriers such as the blood–placenta barrier and BBB (Choi and Zheng, 2009) and modify the process slightly.

Manganese, Molybdenum, and Cobalt

Manganese absorption takes place not only in the GI tract but also in the lungs after inhalation or the skin through dermal permeation (Chen et al., 2018). Uptake into enterocytes occurs mainly through active transport processes mediated by DMT1 (Garcia-Aranda et al., 1983). Since DMT1 also transports other divalent cations, the presence of other metals will lead to competition and will affect manganese import. Besides, the zinc transporter ZIP14 is a manganese transporter that is expressed in the GI tract and that may increase intracellular manganese mostly from the basal direction (Scheiber et al., 2019).

The most commonly oxidized states of manganese are Mn^{2+} and Mn^{3+}. However, Mn^{3+} can increase oxidative stress. Therefore, it is bound by proteins, while the more stable Mn^{2+} may also be found as a free ion in cells. However, the exact nature of the intracellular enterocytic manganese buffering proteins still needs to be determined.

The exact ways of manganese release are currently not well investigated. Possible manganese exporters may be ZnT10 (Quadri et al., 2012), ferroportin (Choi et al., 2019), and SPCA1 (Leitch et al., 2011). Once released into the blood circulation, manganese can be taken up by erythrocytes or transported in serum, where Mn^{2+} is predominantly bound to albumin, while almost all Mn^{3+} is bound to Tf (Chen et al., 2018). Again, the liver is the primary organ involved in the regulation of manganese levels.

Molybdenum is utilized by only three proteins (enzymes) in the human body (sulfite oxidase, xanthine oxidase, and aldehyde oxidase). The mechanisms of absorption are not well investigated. Molybdenum is absorbed in the stomach and gut. Given that DMT1 shows an affinity for molybdenum, the transporter may play a role in the uptake of the metal. High levels of molybdenum have been reported in the liver (Momcilović, 1999), which hints at the liver as a store for the trace metal as discussed for other metals before.

Similar to molybdenum, little is known about the molecular mechanisms of cobalt absorption and intracellular transport. DMT1 also has an affinity for cobalt (Garrick et al., 2003) and may take up the metal into enterocytes. Simultaneous and competitive iron and cobalt absorption was observed (Thomson et al., 1971). This competition seems to result from the interference of cobalt with the iron release on the basal side (Becker et al., 1979). However, foods high in vitamin B12 seem to be the only source of cobalt used by the body. Vitamin B12 is absorbed in the small intestine (ileum) in a two-step process in which vitamin B12 first binds to its receptor on the apical side of enterocytes, followed by a transfer of B12 from its receptor into the cell (Beesley, 1984).

Few data have emerged regarding the mechanisms that control manganese, molybdenum, and cobalt transport across the placenta and BBB. In the placenta, manganese is shuttled from mother to fetus through active transport (Yoon et al., 2009). However, not much is known about how this process is regulated. Molybdenum and cobalt are also able to cross the placenta (Rudge et al., 2009), but the detailed mechanisms involved need to be elucidated in the future.

Tf/TfR processes or DMT1 may mediate the movement of manganese across the BBB. However, most active transport into the brain seems to occur via a calcium uptake mechanism using a calcium concentration-dependent operated calcium channel (Yokel, 2009). Besides some active transport across the BBB, the blood–cerebrospinal fluid (CSF) barrier seems to be the primary route for manganese uptake into the CNS (Bornhorst et al., 2012). However, manganese may also enter the brain via the olfactory tract.

Molybdenum can be detected in the brain. However, the mechanisms of uptake are unclear. Increasing concentrations of molybdenum in the diet of rats lead to a linear increase in brain molybdenum levels. However, beyond 200 μg/kg dietary molybdenum, no further increase in brain molybdenum concentration was observed. This hints at active and controlled mechanisms of transport of molybdenum into the CNS (Wang et al., 1992). One possibility may be uptake by DMT1, which is also proposed for the transport of cobalt into the brain (Zheng et al., 2003). Alternatively, cobalt–Tf complexes may enter the brain after Tf/TfR-mediated transcytosis (Smith et al., 1997).

Not all proteins that can mediate the transport of trace metals are expressed or active in the GI tract. Different tissues utilize a specific set of transport proteins (summarized in Table 6.1). Especially the occurrence of 24 different zinc transporters makes the homeostatic regulation of zinc relatively complicated compared with other trace metals. In addition, zinc has a very high number of different intracellular and extracellular binding proteins, some of which may also play a secondary role in its transport. For example, while most iron in the body is bound to ferritin, Tf, and hemoglobin, zinc can be bound to approximately 2800 different proteins (Andreini et al., 2006), among them enzymes, storage proteins, transcription factors, and replication proteins (Coleman, 1992).

Table 6.1 Overview of proteins involved in the transport of trace metals.

Metal	Transport protein
Fe	DMT1, Nramp1, Nramp2, Fpn1, ceruloplasmin, Tf/TfR1, ferritin, hemoglobin
Zn	Zip1-14, ZnT1-10, MT1-4, Ctr1, calprotectin, Tf, a2-macroglobulin, albumin
Cu	DMT1, Ctr1, Ctr2, CCS, Hah1p, cox17p, ATP7A, ATP7B, cp, Atox1
Mn	DMT1, Zip14, choline transporter, Tf, Ca channel
Pb	DMT1, Ca channels, hemoglobin
Cd	DMT1, Ca channels, Tf, albumin, citrate
Mo	DMT1, Tf, albumin
Co	DMT1, Tf, Fpn1

The elimination of essential trace metals occurs mainly through urine and feces. Interestingly, there is no active mechanism for iron excretion (Kohgo et al., 2008). Iron is recycled after red blood cells (RBCs) are degraded. The iron released from hemoglobin of the RBCs is stored in ferritin/hemosiderin and reutilized. After its release into the blood circulation via Fpn, iron binds to Tf and is transported to the bone marrow for the generation of new hemoglobin and RBCs. Fluctuations in iron levels are buffered by iron storage in the liver, where excess iron is found bound to ferritin and hemosiderin (Kohgo et al., 2008).

Zinc is eliminated primarily through fecal excretion (Krebs, 2000). However, reabsorption of zinc may be an important factor for zinc homeostasis. This reabsorption may occur distally of the main site of absorption in the small intestine (Krebs, 2000).

Excess copper is excreted into bile, but may also be eliminated through urine in situations of high copper burden (Gray et al., 2012). Bile is produced continuously by the liver and stored and concentrated in the gallbladder, before being released into the duodenum. Manganese is also conjugated to bile by the liver and subsequently excreted in feces (Davis et al., 1993), whereas molybdenum and cobalt are regulated by urinary excretion (Turnlund et al., 1995; Mosconi et al., 1994).

Nutritional trace metal needs to maintain an adequate metal status

Besides intact mechanisms for the absorption of trace metals in the GI tract, to maintain adequate trace metal levels, dietary sources must deliver certain quantities of trace metals. The recommended daily allowances (RDAs) and adequate intakes (AIs) for the essential trace metals are shown in Table 6.2. RDAs are evidence-based values, whereas AIs are considered the best estimate of intakes assumed to be adequate. Daily values (DVs) are set by the US Food and Drug Administration (FDA), whereas the RDAs are set by the National Academies of Sciences, Engineering, and Medicine (NDA). DVs are reference amounts of trace metals to consume each day.

However, individual needs may be different due to age, health status, lifestyle factors, or pregnancy. For example, for zinc, it is important to consider the present zinc status of the individual, as the current state affects zinc absorption and excretion. Due to regulatory processes affecting ZIP4 expression and localization at the apical side of the enterocyte, an individual can adjust to a low zinc intake (Andrews, 2008). However, the requirements of such an individual are different compared with an individual who is not adapted to low zinc availability. Pregnancy is another important factor. For example, to ensure there is sufficient daily intake of zinc, the RDA must be increased by +5−10 mg for pregnant and lactating women, which is reflected in the RDA and DVs.

More important than the levels in the diet is the bioavailability of the specific trace metal. For example, from Table 6.2, it can be seen that the DV for manganese for an adult male is 2.3 mg/day, and the DV for zinc is 11 mg/day. Thus, the DV for zinc is 4.8 times greater. However, the body content of manganese and zinc of a 70-kg adult male is 12 and 2300 mg, respectively. Thus, the body contains 191 times the amount of zinc. Therefore, the absorption efficacy determining the bioavailability must be dramatically different between the two trace metals. Indeed, in adults, only about 3%−5% of ingested manganese is absorbed in the GI tract (Chen et al., 2018).

Several genetic and nongenetic factors can influence the bioavailability of a trace metal. Genetic factors may be mutations rendering the proteins necessary for absorption nonfunctional or decrease their efficacy. For example, in humans, mutations in the major zinc uptake transporter ZIP4 cause severe zinc deficiency, as they dramatically reduce the bioavailability of zinc. The associated disease, *acrodermatitis enteropathica (AE)*, is a rare, but if left untreated, lethal autosomal recessive genetic disorder (Küry et al., 2002). Individuals with AE, therefore, must compensate for the decreased efficacy of zinc uptake by supplementing their diet with far more zinc as the RDA (RDA and DV values are calculated for healthy individuals).

Table 6.2 Recommended daily allowances (RDAs) and adequate intakes (AIs), as well as upper tolerable intake levels (ULs) based on the RDAs published by the NDA.

		Metal					
Age	**Intake**	**Iron (mg)**	**Zinc (mg)**	**Copper (µg)**	**Manganese (µg)**	**Molybdenum (µg)**	**Chromium (µg)**
0–6 months							
	RDA/AI*	0.27*	2*	200	3*	2*	0.2*
	Daily value (DV)	11	3	200	600	3	5.5
	Upper limit (UL)	40	4	ND	ND	ND	ND
6–12 months							
	RDA/AI*	11	3	220	600*	3*	5.5*
	DV	11	3	220	600	3	5.5
	UL	40	5	ND	ND	ND	ND
1–3 years							
	RDA/AI*	7	3	340	1200*	17	11*
	DV	7	3	300	1200	17	11
	UL	40	7	1000	2000	300	ND
4–8 years							
	RDA/AI*	10	5	440	1500*	22	15*
	DV	18	11	900	2300	45	35
	UL	40	12	3000	3000	600	ND
9–13 years							
	RDA/AI*	8	8	700	1900* (m) 1600* (f)	34	25 (m) 21 (f)
	DV	18	11	900	2300	45	35
	UL	40	23	5000	6000	1100	ND

Continued

Table 6.2 Recommended daily allowances (RDAs) and adequate intakes (AIs), as well as upper tolerable intake levels (ULs) based on the RDAs published by the NDA.—cont'd

Age	Intake	Metal: Iron (mg)	Zinc (mg)	Copper (μg)	Manganese (μg)	Molybdenum (μg)	Chromium (μg)
14–18 years							
	RDA/AI*	11 (m) 15 (f) 27 (preg) 10 (lact)	11 (m) 9 (f) 12 (preg) 13 (lact)	890 1000 (preg) 1300 (lact)	2200* (m) 1600* (f)	43 50 (preg, lact)	35 (m) 24 (f) 29 (preg) 44 (lact)
	DV	18 27 (preg, lact)	11 13 (preg, lact)	900 1300 (preg, lact)	2300 2600 (preg, lact)	45 50 (preg, lact)	35 44 (preg, lact)
	UL	45	34	8000	9000	1700	
Adult							
	RDA/AI*	8 (m) 18 (f 19–50) 8 (f 50+) 27 (preg) 9 (lact)	11 (m) 8 (f) 11 (preg) 12 (lact)	900 1000 (preg) 1300 (lact)	2300* (m) 1800* (f) 2000 (preg) 2600 (lact)	45 50 (preg, lact)	35* (m 19–50) 24* (f 19–50) 30* (m 50+) 20* (f 50+) 30* (preg) 45* (lact)
	DV	18 27 (preg, lact)	11 13 (preg, lact)	900 1300 (preg, lact)	2300 2600 (preg, lact)	45 50 (preg, lact)	35 45 (preg, lact)
	UL	45	40	10,000	11,000	2000	ND

*, Adequate Intake (AI); *ND*, not determined; *m*, male; *f*, female; *lact*, lactating (breastfeeding); *preg*—pregnant. An RDA/AI for cobalt has not been established yet.

Similarly, the copper bioavailability can be significantly reduced as a result of mutations in the gene coding for ATP7A. The emerging clinical picture is called Menkes disease (Tümer and Møller, 2010). Individuals with Menkes disease need daily injections of copper to bypass the affected GI absorption mechanisms. Apart from these extreme forms of altered bioavailability, genetic variants in metal transport and buffering proteins are mostly responsible for minor interindividual differences.

The most important nongenetic factors affecting the bioavailability of trace metals are other dietary components. Trace metals are normally ingested as part of food that contains several other substances that can interact with the metal or its transporters. For example, the bioavailability of zinc taken up as part of a typical western diet is about 20%–30% (Gallaher et al., 1998), because various dietary compounds present in the diet inhibit the absorption of zinc. Therefore, in the United States, the RDA value is based on a mixed diet with an estimated 20% zinc absorption, whereas in the United Kingdom, a 30% absorption rate is used to determine the RDA.

Among the dietary compounds that inhibit the absorption of zinc are phytates. In fact, phytic acid is the principal dietary factor that limits zinc bioavailability by binding zinc in the gastrointestinal tract (Hambidge et al., 2011). Thereby, the bioavailability is reduced since zinc bound within the phytate–zinc complex cannot be transported through ZIP proteins (Lönnerdal, 2000; Sandstead and Freeland-Graves, 2014). Phytate within food is composed of a mixture of various phosphorylated forms of inositol phosphate (Sandberg and Ahderinne, 1986). The major form is hexaphosphate, but in addition, pentaphosphates, tetraphosphates, and triphosphates are also present. The inhibitory effects of phytic acid on zinc can be predicted by the molar ratios of phytate:zinc in the diet. High phytate:zinc ratios in the diet have been linked with suboptimal zinc status in humans (Roohani et al., 2013). Molar ratios higher than 18:1 (Allen et al., 2006) were deemed to significantly inhibiting zinc absorption. Therefore, dietary sources that contain high zinc and low phytate levels such as meat from various sources (fish, mussels, beef, etc.) deliver zinc with high bioavailability, whereas foods such as refined cereal products, fruit, and vegetables that are high in phytates deliver zinc with low bioavailability. Furthermore, high dietary fiber consumption was reported to have a negative effect on zinc absorption (Knudsen et al., 1996). However, fiber-containing foods usually also contain phytates, which may be, in part, responsible for the effect. Additionally, folic acid, which is commonly prescribed as a supplement during pregnancy, has been shown to decrease the bioavailability of zinc (Ghishan et al., 1986; Simmer et al., 1987; Krebs, 2000).

However, it is the presence of other metals in the diet that may have a major influence on the bioavailability of a certain metal. As discussed earlier, uptake mechanisms for metals may have shared components such as the DMT1, and metals with different affinity and different concentrations in the diet may compete for binding to DMT1 and transport into the enterocyte. Similarly, MTs are intracellular proteins that may bind several different metals such as zinc, copper, and cadmium. Copper and zinc both have competing roles within the body, and an overabundance of copper may lead to zinc deficiency and vice versa (Hill and Matrone, 1970; Mills, 1985; Hall et al., 1979; Oestreicher and Cousins, 1985). Moreover, calcium and iron can interfere with the absorption process for zinc (Wood and Zheng, 1997; Whiting and Wood, 1997; Argiratos and Samman, 1994; Fischer Walker et al., 2005) Similarly, zinc and manganese may interfere with iron absorption (Rossander-Hultén et al., 1991). However, the interactions of trace metals are complex, dynamic, and concentration dependent, and therefore hard to predict.

To account for the various degrees of bioavailability, in 1996, the World Health Organization (WHO) proposed to adjust the daily intake of zinc for both adult men and women based on their diet (with high, medium, or low zinc bioavailability). For example, for an adult male, the RDA for a diet with high zinc bioavailability was 5.6 mg; for medium zinc bioavailability, it was 9.4 mg; and for low zinc bioavailability, it was 18.7 mg. For a female, the RDA for a diet with high zinc bioavailability was 6 mg; for medium zinc bioavailability, it was 10 mg; and for low zinc bioavailability, it was 20 mg.

Taken together, while it is relatively easy to establish RDA and DVs on a population level, it is challenging to evaluate the needs for an individual as many factors such as genetic background, nutrition, health status, age, pregnancy, and so on will have to be taken into account. Especially the higher demand for all essential trace metals during pregnancy makes pregnancy, the phase of brain development of the fetus, a phase where the likelihood for trace metal deficiencies is the highest.

Nutrition and autism spectrum disorder

Individuals with ASD have frequently been reported to have food aversions and sensitivities. In addition, the ASD-associated behavioral issues may impair adequate food intake or result in extreme food selectivity and ritualistic eating behaviors that may facilitate nutritional deficiencies (Sharp et al., 2013). Therefore, inadequate nutrition is more common among children with ASD (Sharp et al., 2013). Also, chronic eating problems may occur (Rastam and Wentz, 2014).

In particular, "pica" may increase the risk of exposure to toxic metals. Pica is an eating disorder that involves eating nonfood items that do not contain a significant nutritional value such as hair, dirt, paint chips, glue, cigarette ashes, and so on (Matson et al., 2013). Pica is a comorbidity observed in individuals with ASD (Neumeyer et al., 2019). Several of the more frequently eaten items, such as dirt or flakes of paint, may contain toxic metals (e.g., lead in lead-based paint).

However, it is very likely that nutritional differences, although possibly a significant modulator of the ASD phenotype, are not a causative factor for ASD. ASD is a neurodevelopmental disorder with onset very early in life, most likely already during brain development in utero. Thus, the maternal nutritional status may be more critical for the etiology of ASD. Nevertheless, the influence of dietary factors on the ASD behavioral phenotype is an important finding, as it represents a possible therapeutic strategy. For example, some reports show that ASD features improve after the introduction of a gluten-free/casein-free diet, fatty acid supplementation, and/or treatment with pre/probiotics (Ng et al., 2019; Hartman and Patel, 2020). However, more research is needed in the future as the results are mixed, and other studies reported no effects (Piwowarczyk et al., 2020).

Few studies have looked at the effects of trace metal supplementation (see Chapter 7), with some of these reporting beneficial effects. However, mostly, the target of nutritional interventions is to improve a GI pathology (that only occurs in a subset of individuals with ASD). Through altered gut–brain signaling, effects on the GI pathology may ultimately translate into alterations in brain function. Trace metal supplementation, in particular, is not expected to have direct effects on the CNS as the BBB will limit the movement of metals into the brain tissue. For example, zinc supplementation will hardly affect brain zinc levels in individuals with normal zinc status. The BBB is a protective structure that allows only minor zinc to transition into the brain if physiological CNS zinc levels are reached. In a study, intravenous delivery of 30 mg zinc acetate for 30 min, followed by continuous delivery of 0.125 mg zinc per minute for 7 hours, raised the plasma zinc concentration approximately by 10-fold in animal studies. However, the zinc concentration in CSF was not altered (Blair-West et al., 1990).

Therefore, a much more promising approach with regard to trace metal supplementation may be not to treat the individual with ASD using commonly available supplements that do not cross the BBB but to avoid maternal trace metal imbalances during pregnancy as a prevention strategy.

Nutritional disorders and autism spectrum disorder

Although zinc is relatively abundant in nature, a plethora of data show that zinc deficiency is one of the most prevalent trace metal deficiencies worldwide (Brown et al., 2004). Recent studies estimated that the prevalence of inadequate dietary zinc intake in adolescent individuals is around 20% worldwide (Willoughby and Bowen, 2014) and that approximately 17.3% of the world's population is at risk or is already zinc deficient (Wessells and Brown, 2012). Another study estimated that one-third to one-half of the world's population is at high risk of a low dietary intake of zinc and risk of developing zinc deficiency (Brown et al., 2001). In particular, given the increased demand during pregnancy, transient zinc deficiency during pregnancy may not be uncommon in humans (Sandstead, 1991, 1995). Indeed, some studies report the prevalence of zinc deficiency among pregnant women to be as high as 57.4% in some regions (Kumera et al., 2015). However, as outlined in the previous chapter, zinc deficiency is complicated to measure, as serum zinc levels can be normal in states of mild deficiency (Bales et al., 1994), and many cases of mild zinc deficiencies will remain undetected.

Zinc deficiency may be responsible for as much as 1% of all deaths globally and 4.4% of deaths in children aged 6 months to 5 years (Fischer Walker et al., 2009). Subjects with severe zinc deficiency display, for example, skin lesions, weight loss, growth retardation, a high susceptibility toward infectious diseases, and interestingly, symptoms of neuropsychiatric disorders (Prasad, 2013; Petrilli et al., 2017). However, these are symptoms of acute zinc deficiency. It is extremely difficult to understand the consequences of zinc deficiency during pregnancy, the critical phase of brain development. After birth, zinc levels may quickly normalize, for example, through being formula-fed versus breastfed. A breastmilk formula mix usually contains adequate zinc levels. Since individuals exposed to prenatal zinc deficiency may have normal zinc status throughout their life and no biomarker exists that may detect prenatal zinc deficiency later in life, human studies are hindered by the ability to detect those individuals. Thus, controlled animal experiments are necessary to investigate a possible relationship to ASD (see Chapter 8). One exception is *AE*.

AE is an autosomal recessive disease (Küry et al., 2016). Thus, although only the child that inherits a mutated ZIP4 gene from the father and the mother will develop the disorder, it is likely that the mother will have diminished efficacy in zinc absorption, having only one functional ZIP4 copy, which may especially impact zinc status during a phase of increased demand such as pregnancy. Due to two faulty copies of the gene, right after birth with the disconnect from the maternal blood circulation, individuals with AE will quickly develop severe zinc deficiency. Thus, individuals with AE may likely have experienced pre- and perinatal low zinc status. Intriguingly, already in 1976, it was noted that mood disturbances occur in individuals with AE, which are particularly evident in infants and young patients. These disturbances were described as "schizoid," and it was reported that children with AE display avoidance of eye contact (Moynahan, 1976), a feature characteristic for ASD. However, AE is a very rare disorder, and studies assessing the frequency of ASD in individuals with AE have not been performed so far.

Iron deficiency results in anemia. Indeed, iron deficiency is a common problem worldwide. For example, iron deficiency anemia and has a prevalence of around 20% in adolescents (De Andrade Cairo et al., 2014). Individuals who are deficient in iron first show a reduction in ferritin levels. In more severe forms, low hemoglobin levels and a decreased number of RBCs and also a low serum iron level and a reduction in transferrin saturation are detected (De Andrade Cairo et al., 2014).

In general, iron deficiency results from a diet low in iron or with low bioavailability of iron, or as a consequence of increased demand (e.g., during pregnancy). However, there are also genetic causes for iron deficiency, such as iron refractory iron deficiency anemia (IRIDA). IRIDA is caused by mutations in the *TMPRSS6* gene, which disrupts iron regulation (Finberg et al., 2008). It was shown that iron deficiency affects the ASD phenotype of children, increasing the severity of ASD behaviors, and primarily intellectual disability (ID) (Gunes et al., 2017). However, as for zinc, acute iron deficiency is only a modifier, but not trigger for ASD. In contrast, a recent study found that anemia diagnosed in early pregnancy ($\leq$30 weeks) is significantly associated as a possible causative factor with an increased risk for the offspring to develop ASD (and attention-deficit/hyperactivity disorder [ADHD], and ID) (Wiegersma et al., 2019).

Copper deficiency caused by low nutritional levels or bioavailability is rare. Copper deficiency leads to a variety of symptoms. The most severe clinical symptoms are hypochromic anemia and peripheral neuropathy (Prohaska, 2014). However, copper deficiency has not been linked to ASD. In contrast, several studies have shown copper overload to be associated with ASD. Excessive copper levels may result from zinc deficiency and may also be caused by genetic factors.

For copper, genetic disorders characterized by both deficiency and overload are known. Genetic copper deficiency occurs in Menkes disease. Menkes disease is an X-chromosomal recessive disorder caused by mutations in *ATP7A*. These mutations lead to abnormal copper transport, ultimately resulting in copper deficiency (Tümer and Møller, 2010). Individuals with Menkes disease may have seizures, hypotonia, abnormal hair, failure to gain weight and growth delay, deterioration of the nervous system, and urologic abnormalities (Caicedo-Herrera et al., 2018). However, Menkes disease has not been linked to ASD.

Wilson's disease, in contrast, is an autosomal recessive disorder caused by mutations in the *ATP7B* gene. ATP7B mediates the export of excess copper from the liver to the bile for excretion. Disruption of this process leads to copper accumulation in various tissues, in particular in the liver and the brain (Członkowska et al., 2018). Young individuals with Wilson's disease show signs of liver disease (jaundice, hepatitis, hepatomegaly/splenomegaly, etc.) and subtle neurological disease. Older individuals mainly display features of neurological or psychiatric diseases, even in the absence of liver disease (Fernando et al., 2020). Among the neurological symptoms are features of movement disorders such as ataxia, dystonia, and stereotypes, features of cognitive impairments such as impaired executive function, learning and memory, and attention, and features of psychiatric diseases such as mood disorders (depression, anxiety, impulsive and aggressive behavior) and psychosis and schizophrenia, as well as seizures and headaches (Fernando et al., 2020).

It can be hypothesized that individuals with Wilson's disease comprise an at-risk group for exposure to higher copper levels in utero and the perinatal period due to the mother being at least heterozygous for the ATP7B mutation. However, copper levels build up only slowly in children with Wilson's disease (first symptoms appear around 3—5 years of age), and despite some of the neurological features shared with ASD, no clear link has emerged yet. Besides, Wilson's disease can be managed by treatments lowering copper levels such as zinc supplementation (again exploiting the antagonistic relationship between zinc and copper) and copper chelators (Aggarwal and Bhatt, 2018). The incidence rate of ASD in children from untreated mothers with Wilson's disease has not been assessed.

Taken together, some links exist between iron and zinc deficiency diseases and ASD. In all cases, it is a deficiency during pregnancy that seems to be associated with altered neurodevelopment. However, as most studies consider the patients with trace metal deficiency for their research and do not investigate the incidence rate of ASD in their children, epidemiological data on the relationship between trace metal deficiency diseases and ASD are still scarce.

References

Aggarwal, A., Bhatt, M., 2018. Advances in treatment of Wilson disease. Tremor Other Hyperkinet Mov 8, 525.

Argiratos, V., Samman, S., 1994. The effect of calcium carbonate and calcium citrate on the absorption of zinc in healthy female subjects. Eur J Clin Nutr 48 (3), 198–204.

Allen, L., Benoist, B., Dary, O., Hurrell, R., 2006. Zinc, folate and other B vitamins, vitamin C, vitamin D, calcium, selenium and fluoride. In: Allen, L., de Benoist, B., Dary, O., Hurrell, R. (Eds.), Guidelines on Food Fortification with Micronutrients. WHO and FAO, Geneva, pp. 124–134.

Andreini, C., Banci, L., Bertini, I., Rosato, A., 2006. Counting the zinc-proteins encoded in the human genome. J Proteome Res 5 (1), 196–201.

Andrews, G.K., 2008. Regulation and function of Zip4, the acrodermatitis enteropathica gene. Biochem Soc Trans 36 (Pt 6), 1242–1246.

Arredondo, M., Muñoz, P., Mura, C.V., Nùñez, M.T., 2003. DMT1, a physiologically relevant apical Cu1+ transporter of intestinal cells. Am J Physiol Cell Physiol 284 (6), C1525–C1530.

Bales, C.W., DiSilvestro, R.A., Currie, K.L., Plaisted, C.S., Joung, H., Galanos, A.N., Lin, P.H., 1994. Marginal zinc deficiency in older adults: responsiveness of zinc status indicators. J Am Coll Nutr 13 (5), 455–462.

Becker, G., Huebers, H., Rummel, W., 1979. Intestinal absorption of cobalt and iron: mode of interaction and subcellular distribution. Blut 38 (5), 397–406.

Beesley, R.C., 1984. Uptake of vitamin B12 by isolated ileal enterocytes. J Nutr Sci Vitaminol 30 (3), 209–220.

Bertinato, J., Swist, E., Plouffe, L.J., Brooks, S.P., L'abbé, M.R., 2008. Ctr2 is partially localized to the plasma membrane and stimulates copper uptake in COS-7 cells. Biochem J 409 (3), 731–740.

Blair-West, J.R., Denton, D.A., Gibson, A.P., McKinley, M.J., 1990. Opening the blood-brain barrier to zinc. Brain Res 507 (1), 6–10.

Bornhorst, J., Wehe, C.A., Hüwel, S., Karst, U., Galla, H.J., Schwerdtle, T., 2012. Impact of manganese on and transfer across blood-brain and blood-cerebrospinal fluid barrier in vitro. J Biol Chem 287 (21), 17140–17151.

Brown, K.H., Rivera, J.A., Bhutta, Z., Gibson, R.S., King, J.C., Lönnerdal, B., Ruel, M.T., Sandtröm, B., Wasantwisut, E., Hotz, C., 2004. International Zinc Nutrition Consultative Group (IZiNCG) technical document #1. Assessment of the risk of zinc deficiency in populations and options for its control. Nutr Bull 25 (1 Suppl. 2), 99–203.

Brown, K.H., Wuehler, S.E., Peerson, J.M., 2001. The importance of zinc in human nutrition and estimation of the global prevalence of zinc deficiency. Food Nutr Bull 22 (2), 113–125.

Caicedo-Herrera, G., Candelo, E., Pinilla, J., Vidal, A., Cruz, S., Pachajoa, H.M., 2018. Novel ATP7Agene mutation in a patient with Menkes disease. Appl Clin Genet 11, 151–155.

Cetin, I., Berti, C., Mandò, C., Parisi, F., 2011. Placental iron transport and maternal absorption. Ann Nutr Metab 59 (1), 55–58.

Chen, P., Bornhorst, J., Aschner, M., 2018. Manganese metabolism in humans. Front Biosci 23, 1655–1679.

Chesters, J.K., Will, M., 1981. Zinc transport proteins in plasma. Br J Nutr 46 (1), 111–118.
Choi, E.K., Nguyen, T.T., Iwase, S., Seo, Y.A., 2019. Ferroportin disease mutations influence manganese accumulation and cytotoxicity. FASEB J 33 (2), 2228–2240.
Choi, B.S., Zheng, W., 2009. Copper transport to the brain by the blood-brain barrier and blood-CSF barrier. Brain Res 1248, 14–21.
Coleman, J.E., 1992. Zinc proteins: enzymes, storage proteins, transcription factors, and replication proteins. Annu Rev Biochem 61, 897–946.
Cousins, R.J., 2010. Gastrointestinal factors influencing zinc absorption and homeostasis. Int J Vitam Nutr Res 80 (4–5), 243–248.
Cousins, R.J., McMahon, R.J., 2000. Integrative aspects of zinc transporters. J Nutr 130 (5), 1384–1387.
Coyle, P., Philcox, J.C., Carey, L.C., Rofe, A.M., 2002. Metallothionein: the multipurpose protein. Cell Mol Life Sci 59 (4), 627–647.
Członkowska, A., Litwin, T., Dusek, P., Ferenci, P., Lutsenko, S., Medici, V., Rybakowski, J.K., Weiss, K.H., Schilsky, M.L., 2018. Wilson disease. Nat Rev Dis Primers 4 (1), 21.
Davis, C.D., Zech, L., Greger, J.L., 1993. Manganese metabolism in rats: an improved methodology for assessing gut endogenous losses. Proc Soc Exp Biol Med 202 (1), 103–108.
De Andrade Cairo, R.C., Rodrigues Silva, L., Carneiro Bustani, N., Ferreira Marques, C.D., 2014. Iron deficiency anemia in adolescents; a literature review. Nutr Hosp 29 (6), 1240–1249.
Fernando, M., van Mourik, I., Wassmer, E., Kelly, D., 2020. Wilson disease in children and adolescents. Arch Dis Child (in press).
Finberg, K.E., Heeney, M.M., Campagna, D.R., Aydinok, Y., Pearson, H.A., Hartman, K.R., Mayo, M.M., Samuel, S.M., Strouse, J.J., Markianos, K., Andrews, N.C., Fleming, M.D., 2008. Mutations in TMPRSS6 cause iron-refractory iron deficiency anemia (IRIDA). Nat Genet 40 (5), 569–571.
Fischer Walker, C.L., Ezzati, M., Black, R.E., 2009. Global and regional child mortality and burden of disease attributable to zinc deficiency. Eur J Clin Nutr 63 (5), 591–597.
Fischer Walker, C., Kordas, K., Stoltzfus, R.J., Black, R.E., 2005. Interactive effects of iron and zinc on biochemical and functional outcomes in supplementation trials. Am J Clin Nutr 82 (1), 5–12.
Gallaher, D.D., Johnson, P.E., Hunt, J.R., Lykken, G.I., Marchello, M.J., 1988. Bioavailability in humans of zinc from beef: intrinsic vs extrinsic labels. Am J Clin Nutr 48 (2), 350–354.
Garcia-Aranda, J.A., Wapnir, R.A., Lifshitz, F., 1983. In vivo intestinal absorption of manganese in the rat. J Nutr 113 (12), 2601–2607.
Garrick, M.D., 2011. Human iron transporters. Genes Nutr 6 (1), 45–54.
Garrick, M.D., Dolan, K.G., Horbinski, C., Ghio, A.J., Higgins, D., Porubcin, M., Moore, E.G., Hainsworth, L.N., Umbreit, J.N., Conrad, M.E., Feng, L., Lis, A., Roth, J.A., Singleton, S., Garrick, L.M., 2003. DMT1: a mammalian transporter for multiple metals. Biometals 16 (1), 41–54.
Garrick, M.D., Garrick, L.M., 2009. Cellular iron transport. Biochim Biophys Acta 1790 (5), 309–325.
Goldani, A.A., Downs, S.R., Widjaja, F., Lawton, B., Hendren, R.L., 2014. Biomarkers in autism. Front Psychiatr 5, 100.
Ghishan, F.K., Said, H.M., Wilson, P.C., Murrell, J.E., Greene, H.L., 1986. Intestinal transport of zinc and folic acid: a mutual inhibitory effect. Am J Clin Nutr 43 (2), 258–262.
Gray, L.W., Peng, F., Molloy, S.A., Pendyala, V.S., Muchenditsi, A., Muzik, O., Lee, J., Kaplan, J.H., Lutsenko, S., 2012. Urinary copper elevation in a mouse model of Wilson's disease is a regulated process to specifically decrease the hepatic copper load. PLoS One 7 (6), e38327.
Gunes, S., Ekinci, O., Celik, T., 2017. Iron deficiency parameters in autism spectrum disorder: clinical correlates and associated factors. Ital J Pediatr 43 (1), 86.

Hall, A.C., Young, B.W., Bremner, I., 1979. Intestinal metallothionein and the mutual antagonism between copper and zinc in the rat. J Inorg Biochem 11 (1), 57–66.
Hambidge, K.M., Miller, L.V., Krebs, N.F., 2011. Physiological requirements for zinc. Int J Vitam Nutr Res 81 (1), 72–78.
Hamza, I., Prohaska, J., Gitlin, J.D., 2003. Essential role for Atox1 in the copper-mediated intracellular trafficking of the Menkes ATPase. Proc Natl Acad Sci USA 100 (3), 1215–1220.
Hartman, R.F., Patel, D., 2020. Dietary approaches to the management of autism spectrum disorders. Adv Neurobiol 24, 547–571.
Hill, C.H., Matrone, G., 1970. Chemical parameters in the study of in vivo and in vitro interactions of transition elements. Fed Proc 29 (4), 1474–1481.
Knudsen, E., Sandström, B., Solgaard, P., 1996. Zinc, copper and magnesium absorption from a fibre-rich diet. J Trace Elem Med Biol 10 (2), 68–76.
Kohgo, Y., Ikuta, K., Ohtake, T., Torimoto, Y., Kato, J., 2008. Body iron metabolism and pathophysiology of iron overload. Int J Hematol 88 (1), 7–15.
Krebs, N.F., 2000. Overview of zinc absorption and excretion in the human gastrointestinal tract. J Nutr 130 (5 Suppl. l), 1374–1377.
Kumera, G., Awoke, T., Melese, T., Esethie, S., Mekuria, G., Mekonnen, F., Ewunetu, T., Gedle, D., 2015. Prevalence of zinc deficiency and its association with dietary, serum albumin and intestinal parasitic infection among pregnant women attending antenatal care at the University of Gondar Hospital, Gondar, Northwest Ethiopia. BMC Nutr 1 (31), 1–11.
Küry, S., Dréno, B., Bézieau, S., Giraudet, S., Kharfi, M., Kamoun, R., Moisan, J.P., 2002. Identification of SLC39A4, a gene involved in acrodermatitis enteropathica. Nat Genet 31 (3), 239–240.
Küry, S., Kharfi, M., Blouin, E., Schmitt, S., Bézieau, S., 2016. Clinical utility gene card for: acrodermatitis enteropathica – update 2015. Eur J Hum Genet 24 (5).
Leitch, S., Feng, M., Muend, S., Braiterman, L.T., Hubbard, A.L., Rao, R., 2011. Vesicular distribution of Secretory Pathway Ca^2+-ATPase isoform 1 and a role in manganese detoxification in liver-derived polarized cells. Biometals 24 (1), 159–170.
Levaot, N., Hershfinkel, M., 2018. How cellular Zn2+signaling drives physiological functions. Cell Calcium 75, 53–63.
Li, Y., Maret, W., 2008. Human metallothionein metallomics. J Anal Atom Spectr 23, 1055–1062.
Lönnerdal, B., 2000. Dietary factors influencing zinc absorption. J Nutr 130 (5 Suppl. l), 1378–1383.
McCarthy, R.C., Kosman, D.J., 2015. Mechanisms and regulation of iron trafficking across the capillary endothelial cells of the blood-brain barrier. Front Mol Neurosci 8, 31.
Maret, W., 2015. Analyzing free zinc(II) ion concentrations in cell biology with fluorescent chelating molecules. Metall 7 (2), 202–211.
Matson, J.L., Hattier, M.A., Belva, B., Matson, M.L., 2013. Pica in persons with developmental disabilities: approaches to treatment. Res Dev Disabil 34 (9), 2564–2571.
Moynahan, E.J., 1976. Letter: zinc deficiency and disturbances of mood and visual behaviour. Lancet 1, 91.
McArdle, H.J., Andersen, H.S., Jones, H., Gambling, L., 2008. Copper and iron transport across the placenta: regulation and interactions. J Neuroendocrinol 20 (4), 427–431.
Mills, C.F., 1985. Dietary interactions involving the trace elements. Annu Rev Nutr 5, 173–193.
Momcilović, B., 1999. A case report of acute human molybdenum toxicity from a dietary molybdenum supplement–a new member of the "Lucor metallicum" family. Arh Hig Rada Toksikol 50 (3), 289–297.
Mosconi, G., Bacis, M., Vitali, M.T., Leghissa, P., Sabbioni, E., 1994. Cobalt excretion in urine: results of a study on workers producing diamond grinding tools and on a control group. Sci Total Environ 150 (1–3), 133–139.

Neumeyer, A.M., Anixt, J., Chan, J., Perrin, J.M., Murray, D., Coury, D.L., Bennett, A., Farmer, J., Parker, R.A., 2019. Identifying associations among Co-occurring medical conditions in children with autism spectrum disorders. Acad Pediatr 19 (3), 300–306.
Ng, Q.X., Loke, W., Venkatanarayanan, N., Lim, D.Y., Soh, A.Y.S., Yeo, W.S., 2019. A systematic review of the role of prebiotics and probiotics in autism spectrum disorders. Medicina 55 (5).
Oestreicher, P., Cousins, R.J., 1985. Copper and zinc absorption in the rat: mechanism of mutual antagonism. J Nutr 115 (2), 159–166.
Paynter, J.A., Grimes, A., Lockhart, P., Mercer, J.F., 1994. Expression of the Menkes gene homologue in mouse tissues lack of effect of copper on the mRNA levels. FEBS Lett 351 (2), 186–190.
Petrilli, M.A., Kranz, T.M., Kleinhaus, K., Joe, P., Getz, M., Johnson, P., Chao, M.V., Malaspina, D., 2017. The emerging role for zinc in depression and psychosis. Front Pharmacol 8, 414.
Piwowarczyk, A., Horvath, A., Pisula, E., Kawa, R., Szajewska, H., 2020. Gluten-free diet in children with autism spectrum disorders: a randomized, controlled, single-blinded trial. J Autism Dev Disord 50 (2), 482–490.
Prasad, A.S., 2013. Discovery of human zinc deficiency: its impact on human health and disease. Adv Nutr 4 (2), 176–190.
Prohaska, J.R., 2014. Impact of copper deficiency in humans. Ann N Y Acad Sci 1314, 1–5.
Quadri, M., Federico, A., Zhao, T., Breedveld, G.J., Battisti, C., Delnooz, C., Severijnen, L.A., Di Toro Mammarella, L., Mignarri, A., Monti, L., Sanna, A., Lu, P., Punzo, F., Cossu, G., Willemsen, R., Rasi, F., Oostra, B.A., van de Warrenburg, B.P., Bonifati, V., 2012. Mutations in SLC30A10 cause parkinsonism and dystonia with hypermanganesemia, polycythemia, and chronic liver disease. Am J Hum Genet 90 (3), 467–477.
Rae, T.D., Schmidt, P.J., Pufahl, R.A., Culotta, V.C., O'Halloran, T.V., 1999. Undetectable intracellular free copper: the requirement of a copper chaperone for superoxide dismutase. Science 284 (5415), 805–808.
Rastam, M., Wentz, E., 2014. ASD, eating problems, and overlap with anorexia and bulimia nervosa. In: Patel, V., Preedy, V., Martin, C. (Eds.), Comprehensive Guide to Autism. Springer, New York, NY, pp. 2015–2034.
Robinson, N.J., Winge, D.R., 2010. Copper metallochaperones. Annu Rev Biochem 79, 537–562.
Roohani, N., Hurrell, R., Kelishadi, R., Schulin, R., 2013. Zinc and its importance for human health: an integrative review. J Res Med Sci 18 (2), 144–157.
Rossander-Hultén, L., Brune, M., Sandström, B., Lönnerdal, B., Hallberg, L., 1991. Competitive inhibition of iron absorption by manganese and zinc in humans. Am J Clin Nutr 54 (1), 152–156.
Rudge, C.V., Röllin, H.B., Nogueira, C.M., Thomassen, Y., Rudge, M.C., Odland, J.Ø., 2009. The placenta as a barrier for toxic and essential elements in paired maternal and cord blood samples of South African delivering women. J Environ Monit 11 (7), 1322–1330.
Sandberg, A.S., Ahderinne, R., 1986. HPLC method for determination of inositol tri-, tetra, penta-, and hexaphosphates in foods and intestinal contents. J Food Sci 51, 547–550.
Sandstead, H.H., 1995. Is zinc deficiency a public health problem? Nutrition 11 (1 Suppl), 87–92.
Sandstead, H.H., 1991. Zinc deficiency. A public health problem? Am J Dis Child 145 (8), 853–859.
Sandstead, H.H., Freeland-Graves, J.H., 2014. Dietary phytate, zinc and hidden zinc deficiency. J Trace Elem Med Biol 28 (4), 414–417.

Sangkhae, V., Nemeth, E., 2019. Placental iron transport: the mechanism and regulatory circuits. Free Radic Biol Med 133, 254–261.
Scheiber, I.F., Wu, Y., Morgan, S.E., Zhao, N., 2019. The intestinal metal transporter ZIP14 maintains systemic manganese homeostasis. J Biol Chem 294 (23), 9147–9160.
Scott, B.J., Bradwell, A.R., 1983. Identification of the serum binding proteins for iron, zinc, cadmium, nickel, and calcium. Clin Chem 29 (4), 629–633.
Sharp, W.G., Berry, R.C., McCracken, C., Nuhu, N.N., Marvel, E., Saulnier, C.A., Klin, A., Jones, W., Jaquess, D.L., 2013. Feeding problems and nutrient intake in children with autism spectrum disorders: a meta-analysis and comprehensive review of the literature. J Autism Dev Disord 43 (9), 2159–2173.
Simmer, K., Iles, C.A., James, C., Thompson, R.P., 1987. Are iron-folate supplements harmful? Am J Clin Nutr 45 (1), 122–125.
Smith, Q.R., Rabin, O., Chikhale, E.G., 1997. Delivery of metals to brain and the role of the blood-brain barrier. In: Connor, J.R. (Ed.), Metals and Oxidative Damage in Neurological Disorders. Springer, Boston, MA, pp. 113–130.
Takeda, A., 2000. Movement of zinc and its functional significance in the brain. Brain Res Brain Res Rev 34 (3), 137–148.
Thomson, A.B., Valberg, L.S., Sinclair, D.G., 1971. Competitive nature of the intestinal transport mechanism for cobalt and iron in the rat. J Clin Invest 50 (11), 2384–2394.
Tümer, Z., Møller, L.B., 2010. Menkes disease. Eur J Hum Genet 18 (5), 511–518.
Turnlund, J.R., Keyes, W.R., Peiffer, G.L., 1995. Molybdenum absorption, excretion, and retention studied with stable isotopes in young men at five intakes of dietary molybdenum. Am J Clin Nutr 62 (4), 790–796.
van den Berghe, P.V., Klomp, L.W., 2009. New developments in the regulation of intestinal copper absorption. Nutr Rev 67 (11), 658–672.
Wang, X., Oberleas, D., Yang, M.T., Yang, S.P., 1992. Molybdenum requirement of female rats. J Nutr 122 (4), 1036–1041.
Wessells, K.R., Brown, K.H., 2012. Estimating the global prevalence of zinc deficiency: results based on zinc availability in national food supplies and the prevalence of stunting. PLoS One 7 (11), e50568.
Whiting, S.J., Wood, R.J., 1997. Adverse effects of high-calcium diets in humans. Nutr Rev 55 (1 Pt 1), 1–9.
Wiegersma, A.M., Dalman, C., Lee, B.K., Karlsson, H., Gardner, R.M., 2019. Association of prenatal maternal anemia with neurodevelopmental disorders. JAMA Psychiatry 1–12.
Willoughby, J.L., Bowen, C.N., 2014. Zinc deficiency and toxicity in pediatric practice. Curr Opin Pediatr 26 (5), 579–584.
Wood, R.J., Zheng, J.J., 1997. High dietary calcium intakes reduce zinc absorption and balance in humans. Am J Clin Nutr 65 (6), 1803–1809.
Yokel, R.A., 2009. Manganese flux across the blood-brain barrier. NeuroMolecular Med 11 (4), 297–310.
Yoon, M., Nong, A., Clewell, H.J., Taylor, M.D., Dorman, D.C., Andersen, M.E., 2009. Evaluating placental transfer and tissue concentrations of manganese in the pregnant rat and fetuses after inhalation exposures with a PBPK model. Toxicol Sci 112 (1), 44–58.
Zheng, W., Aschner, M., Ghersi-Egea, J.F., 2003. Brain barrier systems: a new frontier in metal neurotoxicological research. Toxicol Appl Pharmacol 192 (1), 1–11.

CHAPTER 7

Linking trace metal abnormalities to autism—insights from epidemiological studies

Contents

Lessons from recent meta-analyses and studies investigating trace metal imbalances and autism spectrum disorder

A recent meta-analysis investigating iron levels in individuals with autism spectrum disorder (ASD) compared with individuals without ASD found no significant differences between the two groups (Tseng et al., 2018). In contrast, three meta-analyses investigating zinc levels confirm that low zinc levels are associated with the occurrence of ASD (Sayehmiri et al., 2015; Babaknejad et al., 2016; Saghazadeh et al., 2017). In addition to low zinc levels, the ratio between copper and zinc (Cu/Zn) seems consistently altered, which can be caused by either low zinc, or abnormally high copper, or both. Meta-analyses evaluating the influence of toxic metals further confirmed an association with ASD (Jafari et al., 2017; Saghazadeh and Rezaei, 2017). However, although these studies confirm that trace metal imbalances are indeed linked to the occurrence of ASD, the studies do not imply causation. Therefore, from these results alone, it cannot be concluded that a particular deficiency or overload is causing ASD without performing

Biometals in Autism Spectrum Disorders
ISBN 978-0-12-821132-8
https://doi.org/10.1016/B978-0-12-821132-8.00007-0

further mechanistic studies. For example, given that trace metals show interdependency in our body, from epidemiological studies, it cannot be concluded whether overload with toxic metals is causing an ASD pathology, or whether the overload is a result of less competition for absorption due to zinc deficiency, with zinc deficiency causing the pathology and the toxic metals being irrelevant for ASD. What can be concluded is that an ASD-specific metal imbalance seems to exist, which is characterized by abnormal high Cu/Zn ratio and the presence of increased levels of toxic metals. This characteristic may be caused by different factors, such as the presence of high levels of toxic metals competing with zinc or the lack of zinc, resulting in higher copper levels and an increased abundance of toxic metals. However, likely, other nongenetic and genetic factors may similarly lead to the establishment of such a profile, as immune system abnormalities or stress has been shown to impact on trace metal homeostasis, and both are linked to ASD as well (Grabrucker, 2012).

Furthermore, from large-scale studies that allow investigating metal imbalances in an ASD population across different age groups, a clear correlation between age and the occurrence of imbalances can be observed (Yasuda et al., 2011, 2013). In these studies, the youngest participants showed most alterations. In addition, studies that allow investigating metal levels in utero (Arora et al., 2017) reveal the presence of differences already at this stage of development. Therefore, to understand the influence of biometals and toxic metals on healthy brain development and an association with ASD, more studies aiming at early development during pregnancy and the maternal environment are necessary for the future. So far, few studies have been performed.

Investigating maternal metal imbalances

Measuring the trace metal status of the developing embryo in utero is a difficult task. One of the most suited methods is the measurement of metals in deciduous teeth (milk teeth/primary teeth). To that end, the surface of the dentine is ablated using a laser, and the ejected material is measured using inductively coupled plasma mass spectrometry. Following growth lines such as the neonatal line of the teeth, a particular band of incremental growth lines seen in sections of both enamel and dentine, sampling locations can be selected, which allow a developmental timing. Thus, the history of metal profiles can be reconstructed over the prenatal and early childhood periods (Arora et al., 2017).

This method allows getting some insights into the prenatal period of individuals with ASD. It has been applied several times. Most recently, in an approach that measured teeth from 32 twins and 12 individuals from twin pairs, to limit genetic influences, metal profiles from the non-ASD twins were compared with the siblings with ASD (Arora et al., 2017). The results showed that affected twins had consistently lower manganese levels, both pre- and postnatally. In addition, zinc levels were lower in the affected twin prenatally, but recovered postnatally, even reaching higher levels compared with the nonaffected twin. The concentration of lead was higher in the affected twin during all periods (Arora et al., 2017). Lead and manganese abnormalities were significantly associated with the severity of autistic traits.

Studies investigating trace metal concentrations very early in life show that abnormalities in ASD are present before the onset of visible symptoms, which strengthens the hypothesis that metal dyshomeostasis is contributing as a causative factor to ASD rather than being consequential of ASD. Therefore, brain development in utero is a critical phase for ASD, where the influence of trace metal imbalances may have long-lasting consequences. Unfortunately, few studies investigate specifically the trace metal biology of women during pregnancies.

Due to their increased requirements for zinc, the groups most vulnerable to zinc deficiency include infants, children, the elderly, and also, pregnant and lactating women (Mousa et al., 2019). Due to nutrient demand for fetal growth, childbirth, and producing breast milk during lactation, the requirement of zinc is higher during pregnancy than other life stages. Although the exact daily requirement is not clearly defined, it is acknowledged that zinc intake should be increased for healthy fetal development (Gibson et al., 2016). However, it has been reported that a high number of pregnant women have inadequate zinc intakes (Mousa et al., 2019). For example, it was estimated that pregnant and lactating women consume, on average, only about 9.6 mg of zinc daily, which is well below the recommended intake of 15 mg/day during the second and third trimesters (Parr, 1996).

Epidemiological studies have associated zinc deficiency in pregnancy with impaired immunity, prolonged labor, preterm and postterm births, intrauterine growth retardation (low birth weight [LBW]), and pregnancy-induced hypertension (Caulfield et al., 1998; Ota et al., 2015). Although a recent meta-analysis revealed that the prevalence of ASD is significantly higher in children born preterm (Agrawal et al., 2018) and that LBW is significantly associated with ASD (Lampi et al., 2012), no study so far directly followed a cohort of LBW and/or preterm-born individuals and assessed maternal/prenatal zinc status and the presence of ASD, possibly unifying these risk factors for ASD.

Despite the difficulties to reliably measure trace metal status and in particular zinc status during pregnancy, supplementation with trace minerals and other nutritional components such as folic acid is frequently performed. However, it is possible that mineral and folic acid supplements further contribute to the risk of zinc deficiency by lowering the bioavailability of zinc (Cousins, 2010). For example, it is not uncommon that supplements for pregnant women contain calcium to prevent osteoporosis. Similar to copper, calcium at very high levels may decrease the absorption of zinc (Wood and Zheng, 1997; Whiting and Wood, 1997; Argiratos and Samman, 1994). In addition, some supplements contain iron to prevent anemia. The ingestion of high concentrations of iron might also affect zinc uptake (O'Brien et al., 2000; Fischer Walker et al., 2005). However, while mineral supplements containing calcium, copper, and iron are not standard, folic acid supplementation is generally recommended while trying to conceive and during at least the first 12 weeks of pregnancy. Folic acid is an important nutritional supplement as it has been shown to significantly decrease the risk for the embryo to develop spina bifida. Spina bifida is a neural tube defect that is characterized by a failure of fusion of the caudal neural tube and one of the most common human congenital disabilities (Mitchell et al., 2004). Although the molecular mechanisms underlying the effect of folic acid are largely unknown, folic acid supplementation can prevent the majority of cases of spina bifida. The meta-analysis of 123 studies confirmed that the mandatory folic acid fortification resulted in a lower prevalence of spina bifida (Atta et al., 2016).

However, folic acid has been shown to impact zinc homeostasis. An increase in fecal zinc losses has been observed due to folic acid supplementation, indicating a decreased zinc absorption (Ghishan et al., 1986; Simmer et al., 1987; Krebs, 2000) (Fig. 7.1). Thus, folic acid supplementation may be a risk factor for metal imbalances that are associated with the development of ASD. Indeed, while low levels of folic acid supplementation have shown positive effects on neurodevelopment/ASD in many studies, high (>5 mg/day) folic acid supplementation, resulting in significant amounts of unmetabolized folic acid in serum, was shown to increase the risk for ASD (Gao et al., 2016; Wiens and DeSoto, 2017; Wang et al., 2017). Whether high folic acid serum levels significantly lower serum zinc levels in humans and thereby zinc supply for the developing embryo has not

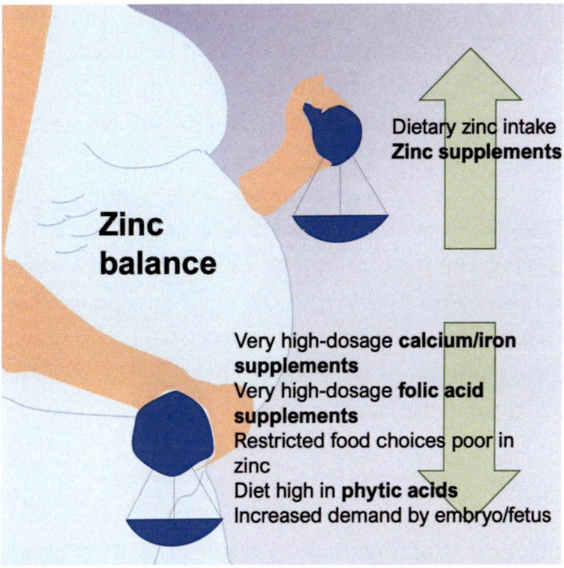

Figure 7.1 Several factors can affect trace metal balance during pregnancy, affecting zinc status in particular. The most common factors include low dietary zinc intake, and the presence of factors lowering the bioavailability such as phytic acid, or the use of folic acid supplementation and mineral supplementation is shown in the figure. However, the kinetics and concentration dependency of the effects have not been well established.

been investigated in detail so far. However, mice fed a diet during pregnancy that contained normal zinc levels but that was supplemented with calcium, iron, folic acid, and phytic acid developed significantly lower blood zinc levels compared with mice on a nonsupplemented diet (Sauer and Grabrucker, 2019).

No general guidelines exist for the supplementation of zinc during pregnancies. Although large-scale epidemiological studies that assess the efficacy of mineral supplementation during pregnancy as a prevention strategy for ASD are missing, several studies investigated supplementation of individuals with ASD as a therapeutic strategy.

Mineral supplementation in autism spectrum disorder

Iron

Few studies investigating iron supplementation in individuals with ASD have been performed so far. A randomized, placebo-controlled trial investigated the effects of ferrous sulfate on sleep abnormalities in individuals with ASD. Impaired sleep is a frequent comorbidity of ASD. However, the study found no significant improvement in the treatment group (Reynolds et al., 2020), in contrast to an open-label treatment pilot trial with oral iron supplementation that showed significantly improved sleep (Dosman et al., 2007).

Another study investigated whether iron supplementation during pregnancy is associated with a reduced risk of ASD. Analyzing 273,107 mother–child pairs, the study found that there was no consistent evidence that iron supplementation is lowering the ASD prevalence (DeVilbiss et al., 2017). Thus, while iron supplementation to correct iron deficiency anemia improves cognitive outcomes (Jáuregui-Lobera, 2014), there is no evidence that prenatal iron supplementation or iron supplementation of individuals with ASD may be beneficial for the ASD pathology.

Zinc

It is known that zinc supplementation can lead to increased absorption of zinc into the blood circulation. Especially individuals with low zinc status respond well to zinc supplementation with an increase in plasma zinc levels (Lowe et al., 2009). However, in many studies, it is difficult to establish whether participants had normal zinc status or whether they were zinc deficient. Especially a mild zinc deficiency may not always result in significantly reduced plasma zinc concentrations. Unfortunately, measuring plasma zinc is the most frequently used assessment of zinc status. Therefore, many studies that have been performed with children with ASD tend not to investigate whether participants with ASD had a low zinc status.

While some studies that investigated the effects of zinc supplementation on children in general have been performed, few studies looked into an ASD population specifically, and no large-scale study that looked at the effects of zinc supplementation of mothers during pregnancy on their children is found, which may be the most promising setup to see effects of zinc on the incidence rate of ASD. However, given that many cases of ASD will have a genetic basis or may be caused by other nongenetic risk factors, a study investigating the effects of preventing zinc deficiency during pregnancy on the prevalence of ASD will require a large number of participants.

The results of human zinc supplementation studies show wide variation, with some of them reporting beneficial effects of zinc supplementation and some not (Shah and Sachdev, 2006). The reason for this inconsistency is mostly likely found in differences in the timing of sampling, laboratory methods, sample sizes, and differences in the underlying zinc nutriture of the participants observed. In general, recent meta-analyses found benefits for young children in infant immunity and indicate that zinc supplementation increases height and weight (Liu et al., 2018; Lassi et al., 2020). However, three meta-analyses revealed no significant impact of zinc supplementation on intelligence, executive function, and motor skills in general (Nissensohn et al., 2013; Warthon-Medina et al., 2015; Gogia and Sachdev, 2012). Nevertheless, some aspects of cognition improved following zinc supplementation in several studies (Warthon-Medina et al., 2015). For example, measures of intelligence (Umamaheswari et al., 2011) and memory (Gewa et al., 2009; Tupe and Chiplonkar, 2009; Umamaheswari et al., 2011) showed improvement with increased zinc intake/levels.

Most of the trials involved only healthy children, and due to the lack of a reliable biomarker of zinc status in humans, it is not known for most studies whether children with zinc deficiency were among the participants. Zinc-deficient children, however, are likely to benefit more, or solely, from normalized zinc levels. One study that performed zinc supplementation of malnourished infants reported improved neuropsychological functions (Bhatnagar and Taneja, 2001). However, other deficiencies due to the general malnourishment may limit the effects of zinc in these children.

Human studies that are based on controlled clinical trials using zinc supplementation in children with ASD so far have rarely been performed. The studies that were conducted found beneficial effects. For example, a studied that included participants with autism and Asperger's syndrome found that after zinc and vitamin B6 supplementation, the severity of symptoms decreased with respect to awareness, receptive language, focus, attention and hyperactivity, eye contact, sound and tactile sensitivity, and seizures, in autistic individuals but not in the group of participants with Asperger's syndrome (Russo and Devito, 2011).

Another study similarly reports improvements concerning hyperactivity and repetitive movements after zinc supplementation. However, this study distinguished between individuals with ASD and gastrointestinal (GI) diseases that frequently occur as a comorbidity in ASD, and children without GI pathology. The improvements were specific to the group of autistic children with GI disease (Russo, 2011). This group may comprise a subpopulation of individuals with ASD that have a compromised zinc absorption due to the GI problems and, therefore, benefit especially from zinc therapy. In line with these results, studies determining the effects of zinc supplementation in children with attention-deficit hyperactivity disorder (ADHD) also found beneficial effects such as an improvement with respect to inattention (Noorazar et al., 2020; Salehi et al., 2016).

In another study, zinc supplementation was reported to increase the cognitive motor performance of children with ASD, as well as to lower the serum levels of copper, thereby normalizing the Cu/Zn ratio that has been consistently shown to be significantly higher in the ASD population (Meguid et al., 2019).

All these studies performed with participants with ASD included only children at an age where a diagnosis of ASD was established. No study so far investigated zinc supplementation of very young children before the symptoms of ASD become evident and followed the participants to investigate whether zinc supplementation delays onset or modifies the severity of symptoms or leads to an overall decrease in the numbers of children diagnosed with ASD in the supplemented group.

Similarly, studies that investigate the influence of zinc supplementation, intake, or status during pregnancy with a focus on the occurrence of neurodevelopmental disorders, like ASD, later in life are almost completely missing. However, it has been reported that low maternal intake of zinc during pregnancy and lactation is associated with less focused attention in the born children (Bhatnagar and Taneja, 2001).

Copper and other trace metals

Increased copper levels have been reported in ASD. Zinc supplementation is an approach that has been shown to normalize copper levels due to the antagonistic relationship of the metals (Russo and Devito, 2011; Meguid et al., 2019). Thus, some of the beneficial effects of zinc supplementation may be achieved by lowering copper levels. Alternatively, copper levels can be reduced through treatment with copper-specific chelators such as penicillamine, trientine, and dimercaprol (Ala et al., 2007). D-Penicillamine was even suggested as an ASD prevention strategy in newborns (Lakatos et al., 2018). However, large-scale randomized and placebo-controlled clinical trials are missing (Sinha et al., 2006). A case study reports a pregnant woman with elevated copper levels who had a child with ASD that also presented high copper levels. During a further pregnancy, the woman received a zinc supplementation therapy that reduced copper levels and led to the birth of a child that did not show elevated copper levels (Walker et al., 2011). The child has not developed signs of ASD so far (Walker, personal communication).

In addition to copper, increased levels of mercury and lead have been reported in ASD. However, a meta-analysis of studies using dimercapto-succinic acid (DMSA), a medication used to treat lead, mercury, and arsenic poisoning, did overall not find this pharmaceutical chelation therapy to be an effective intervention for ASD (James et al., 2015). Taken together, there is no compelling evidence that suggests that metal chelation therapy could be an effective treatment for ASD.

Conclusions

Both epidemiological studies investigating trace metal abnormalities and clinical studies investigating the efficacy of supplementation or chelation suffer from the same limitation: They do not capture the critical time window during development, where the effects of trace metal abnormalities most likely are most potent or occur exclusively. Almost all epidemiological studies measure individuals with ASD and compare them to neurotypical age-matched controls. However, animal studies (see Chapter 9) have shown that altered trace metal status in utero may cause persistent changes that are irreversible, even if a healthy trace metal balance is restored. Therefore, many individuals with ASD in epidemiological studies may no longer show metal imbalances but still may suffer from the consequences. In line with this, studies have repeatedly shown that the older individuals with ASD present fewer metal abnormalities in their biosamples (Yasuda et al., 2011), while metal abnormalities in the group with ASD can be traced back to their development in utero (Arora et al., 2017).

Thus, many metal abnormalities may go undetected in studies using individuals with ASD that are usually at least 2–3 years of age before being diagnosed, while studies investigating large groups of pregnant women or newborns that follow the development of the children and perform a correlation between metal imbalances and the occurrence of ASD are missing.

Similarly, if metal imbalances cause irreversible alterations in brain development in utero, the success of supplementation and chelation therapies must be limited, especially in older individuals with ASD that no longer have trace metal abnormalities. However, no large-scale supplementation or chelation studies have been performed with pregnant women as participants, and follow-up evaluation, whether the treatment affects the incidence rate of ASD has been performed so far. Therefore, many studies are inconclusive and suffer from a focus on the individual with ASD rather than the maternal environment during the critical phase of brain development.

References

Agrawal, S., Rao, S.C., Bulsara, M.K., Patole, S.K., 2018. Prevalence of autism spectrum disorder in preterm infants: a meta-analysis. Pediatrics 142 (3).

Ala, A., Walker, A.P., Ashkan, K., Dooley, J.S., Schilsky, M.L., 2007. Wilson's disease. Lancet 369 (9559), 397–408.

Argiratos, V., Samman, S., 1994. The effect of calcium carbonate and calcium citrate on the absorption of zinc in healthy female subjects. Eur J Clin Nutr 48 (3), 198–204.
Arora, M., Reichenberg, A., Willfors, C., Austin, C., Gennings, C., Berggren, S., Lichtenstein, P., Anckarsäter, H., Tammimies, K., Bölte, S., 2017. Fetal and postnatal metal dysregulation in autism. Nat Commun 8, 15493.
Atta, C.A., Fiest, K.M., Frolkis, A.D., Jette, N., Pringsheim, T., St Germaine-Smith, C., Rajapakse, T., Kaplan, G.G., Metcalfe, A., 2016. Global birth prevalence of spina bifida by folic acid fortification status: a systematic review and meta-analysis. Am J Public Health 106 (1), e24–34.
Babaknejad, N., Sayehmiri, F., Sayehmiri, K., Mohamadkhani, A., Bahrami, S., 2016. The relationship between zinc levels and autism: a systematic review and meta-analysis. Iran J Child Neurol 10 (4), 1–9.
Bhatnagar, S., Taneja, S., 2001. Zinc and cognitive development. Br J Nutr 85 (Suppl. 2), 139–145.
Caulfield, L.E., Zavaleta, N., Shankar, A.H., Merialdi, M., 1998. Potential contribution of maternal zinc supplementation during pregnancy to maternal and child survival. Am J Clin Nutr 68 (2 Suppl.), 499–508.
Cousins, R.J., 2010. Gastrointestinal factors influencing zinc absorption and homeostasis. Int J Vitam Nutr Res 80 (4–5), 243–248.
DeVilbiss, E.A., Magnusson, C., Gardner, R.M., Rai, D., Newschaffer, C.J., Lyall, K., Dalman, C., Lee, B.K., 2017. Antenatal nutritional supplementation and autism spectrum disorders in the Stockholm youth cohort: population based cohort study. BMJ 359, j4273.
Dosman, C.F., Brian, J.A., Drmic, I.E., Senthilselvan, A., Harford, M.M., Smith, R.W., Sharieff, W., Zlotkin, S.H., Moldofsky, H., Roberts, S.W., 2007. Children with autism: effect of iron supplementation on sleep and ferritin. Pediatr Neurol 36 (3), 152–158.
Fischer Walker, C., Kordas, K., Stoltzfus, R.J., Black, R.E., 2005. Interactive effects of iron and zinc on biochemical and functional outcomes in supplementation trials. Am J Clin Nutr 82 (1), 5–12.
Gao, Y., Sheng, C., Xie, R.H., Sun, W., Asztalos, E., Moddemann, D., Zwaigenbaum, L., Walker, M., Wen, S.W., 2016. New perspective on impact of folic acid supplementation during pregnancy on neurodevelopment/autism in the offspring children - a systematic review. PLoS One 11 (11), e0165626.
Gewa, C.A., Weiss, R.E., Bwibo, N.O., Whaley, S., Sigman, M., Murphy, S.P., Harrison, G., Neumann, C.G., 2009. Dietary micronutrients are associated with higher cognitive function gains among primary school children in rural Kenya. Br J Nutr 101 (9), 1378–1387.
Ghishan, F.K., Said, H.M., Wilson, P.C., Murrell, J.E., Greene, H.L., 1986. Intestinal transport of zinc and folic acid: a mutual inhibitory effect. Am J Clin Nutr 43 (2), 258–262.
Gibson, R.S., King, J.C., Lowe, N., 2016. A review of dietary zinc recommendations. Food Nutr Bull 37 (4), 443–460.
Grabrucker, A.M., 2012. Environmental factors in autism. Front Psychiatry 3, 118.
Gogia, S., Sachdev, H.S., 2012. Zinc supplementation for mental and motor development in children. Cochrane Database Syst Rev 12, CD007991.
Jafari, T., Rostampour, N., Fallah, A.A., Hesami, A., 2017. The association between mercury levels and autism spectrum disorders: a systematic review and meta-analysis. J Trace Elem Med Biol 44, 289–297.
James, S., Stevenson, S.W., Silove, N., Williams, K., 2015. Chelation for autism spectrum disorder (ASD). Cochrane Database Syst Rev 5, CD010766.
Jáuregui-Lobera, I., 2014. Iron deficiency and cognitive functions. Neuropsychiatr Dis Treat 10, 2087–2095.

Krebs, N.F., 2000. Overview of zinc absorption and excretion in the human gastrointestinal tract. J Nutr 130 (5S Suppl.), 1374–1377.
Lakatos, L., Pataki, I., Balla, G., Nagy, Z.V., 2018. Potential role of penicillamine in the early diagnosis and prevention of autism spectrum disorder (ASD). SF J Autism 2, 1.
Lampi, K.M., Lehtonen, L., Tran, P.L., Suominen, A., Lehti, V., Banerjee, P.N., Gissler, M., Brown, A.S., Sourander, A., 2012. Risk of autism spectrum disorders in low birth weight and small for gestational age infants. J Pediatr 161 (5), 830–836.
Lassi, Z.S., Kurji, J., Oliveira, C.S., Moin, A., Bhutta, Z.A., 2020. Zinc supplementation for the promotion of growth and prevention of infections in infants less than six months of age. Cochrane Database Syst Rev 4, CD010205.
Liu, E., Pimpin, L., Shulkin, M., Kranz, S., Duggan, C.P., Mozaffarian, D., Fawzi, W.W., 2018. Effect of zinc supplementation on growth outcomes in children under 5 years of age. Nutrients 10 (3).
Lowe, N.M., Fekete, K., Decsi, T., 2009. Methods of assessment of zinc status in humans: a systematic review. Am J Clin Nutr 89 (6), 2040–2051.
Meguid, N.A., Bjørklund, G., Gebril, O.H., Dosa, M.D., Anwar, M., Elsaeid, A., Gaber, A., Chirumbolo, S., 2019. The role of zinc supplementation on the metallothionein system in children with autism spectrum disorder. Acta Neurol Belg 119 (4), 577–583.
Mitchell, L.E., Adzick, N.S., Melchionne, J., Pasquariello, P.S., Sutton, L.N., Whitehead, A.S., 2004. Spina bifida. Lancet 364 (9448), 1885–1895.
Mousa, A., Naqash, A., Lim, S., 2019. Macronutrient and micronutrient intake during pregnancy: an overview of recent evidence. Nutrients 11 (2).
Nissensohn, M., Sánchez-Villegas, A., Fuentes Lugo, D., Henríquez Sánchez, P., Doreste Alonso, J., Skinner, A.L., Medina, M.W., Lowe, N.M., Hall Moran, V., Serra-Majem, L., 2013. Effect of zinc intake on mental and motor development in infants: a meta-analysis. Int J Vitam Nutr Res 83 (4), 203–215.
Noorazar, S.G., Malek, A., Aghaei, S.M., Yasamineh, N., Kalejahi, P., 2020. The efficacy of zinc augmentation in children with attention deficit hyperactivity disorder under treatment with methylphenidate: a randomized controlled trial. Asian J Psychiatr 48, 101868.
O'Brien, K.O., Zavaleta, N., Caulfield, L.E., Wen, J., Abrams, S.A., 2000. Prenatal iron supplements impair zinc absorption in pregnant Peruvian women. J Nutr 130 (9), 2251–2255.
Ota, E., Mori, R., Middleton, P., Tobe-Gai, R., Mahomed, K., Miyazaki, C., Bhutta, Z.A., 2015. Zinc supplementation for improving pregnancy and infant outcome. Cochrane Database Syst Rev (2), CD000230.
Parr, R., 1996. Assessment of dietary intakes. In: World Health Organization. Trace Elements in Human Nutrition and Health; Geneva, Switzerland, pp. 265–288.
Reynolds, A.M., Connolly, H.V., Katz, T., Goldman, S.E., Weiss, S.K., Halbower, A.C., Shui, A.M., Macklin, E.A., Hyman, S.L., Malow, B.A., 2020. Randomized, placebo-controlled trial of ferrous sulfate to treat insomnia in children with autism spectrum disorders. Pediatr Neurol 104, 30–39.
Russo, A.J., 2011. Increased copper in individuals with autism normalizes post zinc therapy more efficiently in individuals with concurrent GI disease. Nutr Metab Insights 4, 49–54.
Russo, A.J., Devito, R., 2011. Analysis of copper and zinc plasma concentration and the efficacy of zinc therapy in individuals with Asperger's syndrome, pervasive developmental disorder not otherwise specified (PDD-NOS) and autism. Biomark Insights 6, 127–133.

Saghazadeh, A., Ahangari, N., Hendi, K., Saleh, F., Rezaei, N., 2017. Status of essential elements in autism spectrum disorder: systematic review and meta-analysis. Rev Neurosci 28 (7), 783–809.
Saghazadeh, A., Rezaei, N., 2017. Systematic review and meta-analysis links autism and toxic metals and highlights the impact of country development status: higher blood and erythrocyte levels for mercury and lead, and higher hair antimony, cadmium, lead, and mercury. Prog Neuropsychopharmacol Biol Psychiatry 79 (Pt B), 340–368.
Salehi, B., Mohammadbeigi, A., Sheykholeslam, H., Moshiri, E., Dorreh, F., 2016. Omega-3 and Zinc supplementation as complementary therapies in children with attention-deficit/hyperactivity disorder. J Res Pharm Pract 5 (1), 22–26.
Sauer, A.K., Grabrucker, A.M., 2019. Zinc deficiency during pregnancy leads to altered microbiome and elevated inflammatory markers in mice. Front Neurosci 13, 1295.
Sayehmiri, F., Babaknejad, N., Bahrami, S., Sayehmiri, K., Darabi, M., Rezaei-Tavirani, M., 2015. Zn/Cu levels in the field of autism disorders: a systematic review and meta-analysis. Iran J Child Neurol 9 (4), 1–9.
Shah, D., Sachdev, H.P., 2006. Zinc deficiency in pregnancy and fetal outcome. Nutr Rev 64 (1), 15–30.
Simmer, K., Iles, C.A., James, C., Thompson, R.P., 1987. Are iron-folate supplements harmful? Am J Clin Nutr 45 (1), 122–125.
Sinha, Y., Silove, N., Williams, K., 2006. Chelation therapy and autism. BMJ 333 (7571), 756.
Tseng, P.T., Cheng, Y.S., Chen, Y.W., Stubbs, B., Whiteley, P., Carvalho, A.F., Li, D.J., Chen, T.Y., Yang, W.C., Tang, C.H., Chu, C.S., Yang, W.C., Liang, H.Y., Wu, C.K., Yen, C.F., Lin, P.Y., 2018. Peripheral iron levels in children with autism spectrum disorders vs controls: a systematic review and meta-analysis. Nutr Res 50, 44–52.
Tupe, R.P., Chiplonkar, S.A., 2009. Zinc supplementation improved cognitive performance and taste acuity in Indian adolescent girls. J Am Coll Nutr 28 (4), 388–396.
Umamaheswari, K., Bhaskaran, M., Krishnamurthy, G., Vasudevan, H., Vasudevan, K., 2011. Effect of iron and zinc deficiency on short term memory in children. Indian Pediatr 48 (4), 289–293.
Walker, L.R., Rattigan, M., Canterino, J., 2011. A case of isolated elevated copper levels during pregnancy. J Pregnancy 2011, 385767.
Wang, M., Li, K., Zhao, D., Li, L., 2017. The association between maternal use of folic acid supplements during pregnancy and risk of autism spectrum disorders in children: a meta-analysis. Mol Autism 8, 51.
Warthon-Medina, M., Moran, V.H., Stammers, A.L., Dillon, S., Qualter, P., Nissensohn, M., Serra-Majem, L., Lowe, N.M., 2015. Zinc intake, status and indices of cognitive function in adults and children: a systematic review and meta-analysis. Eur J Clin Nutr 69 (6), 649–661.
Whiting, S.J., Wood, R.J., 1997. Adverse effects of high-calcium diets in humans. Nutr Rev 55 (1 Pt 1), 1–9.
Wiens, D., DeSoto, M.C., 2017. Is high folic acid intake a risk factor for autism?-A review. Brain Sci 7 (11).
Wood, R.J., Zheng, J.J., 1997. High dietary calcium intakes reduce zinc absorption and balance in humans. Am J Clin Nutr 65 (6), 1803–1809.
Yasuda, H., Kobayashi, M., Yasuda, Y., Tsutsui, T., 2013. Estimation of autistic children by metallomics analysis. Sci Rep 3, 1199.
Yasuda, H., Yoshida, K., Yasuda, Y., Tsutsui, T., 2011. Infantile zinc deficiency: association with autism spectrum disorders. Sci Rep 1, 129.

CHAPTER 8

The specific role of zinc in autism spectrum disorders

Contents

Zinc deficiency as a cause for synaptopathies

A synaptopathy is defined as a disease of the central or peripheral nervous system caused by an impaired or excessive synapse formation and/or synapse dysfunction. Synaptopathies can have different causes such as mutations in essential synaptic proteins regulating synapse maturation, neurotransmitter release and neurotransmitter signaling, and synaptic plasticity, or the presence of nongenetic factors affecting the same processes. Evidence is mounting that autism spectrum disorders (ASDs) are synaptopathies (Won et al., 2013; Guang et al., 2018). An analysis of candidate genes of ASD revealed that the majority of ASD-associated genes that have been identified so far, and that seemingly play diverse roles in chromatin remodeling, metabolism, mRNA translation, and synaptic functions, ultimately converge in common pathways affecting synapse formation and function (Huguet et al., 2013). In addition, animal models for ASD repeatedly display alterations in synapse numbers, abnormal synaptic plasticity, and synaptic dysfunction (Zoghbi and Bear, 2012).

In particular, glutamatergic synapses are in the focus of ASD research (Yang and Chang, 2014). At glutamatergic synapses, the proteins encoded by critical ASD-associated genes such as neurexin *(NRXN)*, neuroligin *(NLGN)*, SH3 and multiple ankyrin repeat domains *(SHANK)*, tuberous sclerosis 1/2 *(TSC1/2)*, fragile X mental retardation 1 *(FMR1)*, phosphatase

Biometals in Autism Spectrum Disorders
ISBN 978-0-12-821132-8
https://doi.org/10.1016/B978-0-12-821132-8.00008-2

and tensin homolog *(PTEN)*, neurofibromatosis type 1 *(NF1)*, and methyl CpG binding protein 2 *(MECP2)* function in common synaptic signaling pathways. The two most important signaling pathways that are also linked by several shared proteins are the NRXN-NLGN-SHANK pathway and the mTOR/PI3K (mammalian target of rapamycin/phosphatidylinositol 3-kinase) pathway (Bourgeron, 2009). Both pathways contain key proteins that are regulated by zinc. Thus, the functionality of the biological pathways is affected by zinc availability and zinc signaling inside cells. The NRXN-NLGN-SHANK pathway has a major role in regulating synaptic spine density, neurotransmitter receptor signaling, morphological changes accompanying synaptic plasticity, and dendrite formation (Chen et al., 2014). The mTOR/PI3K pathway regulates protein translation (including synaptic proteins) and autophagy and is involved in neurogenesis (LiCausi and Hartman, 2018). mTOR signaling is activated in several ways, including MAPK/ERK signaling.

Thus, neurogenesis is also dependent on an intact mitogen-activated protein kinase (MAPK)/ERK pathway. It was shown that MAPK3/ERK1, an ASD candidate gene, is regulated by zinc (Nuttall and Oteiza, 2012). Furthermore, zinc is essential for neurite outgrowth by regulating the expression of genes, including fibroblast growth factor receptor 4 *(FGFR4)*, which is implicated in neurite expansion and growth (Sanford et al., 2019), and regulates microtubule formation (see Chapter 4). Zinc also regulates the myelin basic protein (MBP), a potential candidate autoantigen in ASD (Gonzalez-Gronow et al., 2015).

Taken together, ASD candidate genes converge on pathways that regulate neurogenesis, neuronal migration, neurite growth, and synapse formation and function (Gilbert and Man, 2017). Zinc plays a crucial role in all these processes (see Chapter 4).

Zinc deficiency as a cause for shankopathies

Excitatory glutamatergic synapses harbor a complex postsynaptic network of proteins—the postsynaptic density (PSD). The PSD comprises interacting proteins that are linked to cytoskeletal proteins, postsynaptic neurotransmitter and cell surface receptors, cell adhesion proteins, and proteins and enzymes that transduce and integrate synaptic signals (Boeckers, 2006). Within this network, SHANK proteins occupy a central position. They are dynamically regulated multidomain proteins with a high number of interaction partners and considered to be master scaffolding proteins of the PSD (Boeckers et al., 2002). The family of SHANK genes encodes three members, all of which have been linked to ASD (Leblond et al., 2014).

ASD-associated mutations in SHANK3 were the first to be found and included deletions, duplications, and point mutations (Durand et al., 2007; Moessner et al., 2007; Gauthier et al., 2009). In addition, the loss of one copy of *SHANK3* in humans is associated with Phelan McDermid syndrome (22q13 deletion syndrome) (Costales and Kolevzon, 2015). The syndrome is characterized, among others, by moderate to profound mental retardation, neonatal hypotonia, and ASD-like behavior (Oberman et al., 2015). Subsequently, mutations associated with ASD were reported in *SHANK2* (Berkel et al., 2010; Pinto et al., 2010).

SHANK proteins directly bind to many proteins at the PSD, among them GKAP and LASP1. Moreover, they are linked to the transsynaptic complexes of NLGN and NRXN (Meyer et al., 2004), mediating post- and presynaptic structural plasticity (Arons et al., 2012). They also cluster neurotransmitter receptors, including group I mGluRs (Verpelli et al., 2011), *N*-methyl-D aspartate (NMDA) receptors (R) (via GKAP/PSD95) (Naisbitt et al., 1999), and α-amino-3-hydroxy-5-methyl-4-isoxazole propionic acid (AMPA) receptors (directly or via GRIP1) (Uchino et al., 2006). Mutations/deletions in every protein mentioned earlier are associated with ASD. Thus, SHANK proteins are in the center of an ASD-associated signaling complex at glutamatergic synapses.

In 2002, it was already shown that the PSD contains relatively high amounts of zinc (4.1 nmol/mg protein) (Jan et al., 2002). It was further shown that after in vitro disassembly of PSDs, spontaneous reassembly could only be induced by the addition of zinc, and it was hypothesized that specific zinc-binding proteins mediate this process within the PSD (Jan et al., 2002). In 2006, SHANK3 was identified as a zinc-binding protein (Baron et al., 2006). The fact that SHANK3 can self-associate in the presence of zinc, forming large 2D platforms, makes the protein the best candidate to date for the process described by Jan et al.

SHANK3 and also SHANK2 bind zinc with their C-terminal sterile-α-motif (SAM) domain (Baron et al., 2006). The SHANK3 SAM domain forms a helical polymer, and by SAM domains aligning side by side in an antiparallel manner, a sheetlike structure is formed (Baron et al., 2006). These SHANK3 and most likely SHANK2 and SHANK2/SHANK3 sheets allow the remaining domains of the proteins to be accessible for interactions with other proteins, for example, other PSD proteins and cytoskeletal components (Grabrucker, 2014).

In line with this, it was shown that the SAM domains of SHANK2 and SHANK3 are necessary and sufficient for the synaptic localization of the proteins (Boeckers et al., 2005; Grabrucker et al., 2009), whereas SHANK1, whose SAM domain is nonzinc binding due to differences in the amino acid sequence, is recruited to the PSD via its PDZ domain and thus requires a preformed PSD scaffold (Grabrucker et al., 2011a). Several ASD-associated mutations in *SHANK2* and *SHANK3* lead to premature stop codons and therefore C-terminally truncated proteins that lack the zinc-binding SAM domains and thus no longer localize at PSDs.

Zinc binding to SHANK3 regulates the amount of PSD-bound proteins and the packing density of the SHANK proteins within sheets (Baron et al., 2006; Grabrucker et al., 2011a; Tao-Cheng et al., 2016) through stabilizing salt bridges across two SAM domains by directly binding one amino acid in each SHANK molecule. Thus, the addition of zinc regulates SHANK3 dynamics and leads to an increase in PSD-bound SHANK3 and SHANK2 versus soluble SHANK3/2 within the synaptic spine (Grabrucker et al., 2011a). It was shown that an increase in postsynaptic zinc and PSD-bound SHANK3 occurs in a synaptic activity-dependent manner (Grabrucker et al., 2014). Two sources of zinc may account for the activity-dependent changes in free zinc that is able to bind to SHANK2 and SHANK3. One is the presynaptically released zinc that may enter the postsynapse through NMDA and AMPA receptor channels and calcium channels (see Chapter 4), and the other is postsynaptically released zinc from metallothionein 3 (MT3). The latter seems to play a more important role in the regulation of SHANK3 (Grabrucker et al., 2014). Therefore, the activity and zinc-dependent regulation of SHANK2 and SHANK3 scaffold proteins, and thus of the PSD network, may occur not only at zincergic synapses since MT3 is widely expressed in neurons throughout the brain. However, the presynaptically released zinc from zincergic synapses may provide additional regulatory input.

From this zinc dependency of SHANK2/3, a model that describes a form of synaptic plasticity emerges, which is driven by zinc signaling and which may be critical for healthy brain development: It is possible that SHANK3 sheets quickly change shape by assembly or disassembly of the sheet from smaller units. Synaptic activity, thereby, results in addition and prolonged synaptic inactivity in the removal of SHANK3 interactions at the PSD (Gundelfinger et al., 2006), following the "use it or lose it" principle.

The amount of PSD-bound SHANK2/3, in return, has consequences for synaptic signaling since SHANK2 and SHANK3 function as docking sites for a large number of interacting proteins. For example, the amount of SHANK proteins attached to the PSD scaffold has an impact on proper synapse establishment with both zinc deficiency and SHANK2 and SHANK3 deficiency, resulting in less synaptic contacts that are formed during neuronal development in vitro and in vivo (Grabrucker et al., 2011a, 2014; Schmeisser et al., 2012). The synaptic levels of SHANK3 also have effects on the maturation and "strengthening" of synaptic contacts as part of synaptic plasticity. For example, through NLGN and NRXN, an increase in the postsynaptic scaffold driven by SHANK3 will elicit presynaptic changes, such as an increase in the vesicle pool (Arons et al., 2012). Together, abnormal SHANK function, whether induced by low availability of zinc or mutations in the protein, will affect the overall connectivity of neurons and, thus, brain regions (Fig. 8.1A). Altered connectivity based on synapse dysfunction is central to the ASD pathology (Bourgeron, 2015).

Besides, SHANK proteins modulate neurotransmission. The availability for additional binding sites for receptors and their function might, thus, be dependent on synaptic activity and zinc-dependent increases in PSD-bound SHANK3. For example, it was shown that SHANK2 and SHANK3 regulate AMPAR function by enabling a subunit switch from GluA2-lacking to GluA2-containing AMPARs. GluA2 levels at the synapse increase with an increase in SHANK2 and SHANK3, which is driven by activity and zinc release (Ha et al., 2018).

The functions of SHANK proteins are embedded in signaling pathways that have been identified to play a vital role in ASD through mutations found in individuals with ASD. ASD-associated mutations have been described in many SHANK2 and SHANK3 interacting proteins (Fig. 8.1B). Monogenic forms of ASD have been associated with proteins that are linked to genetic diseases such as tuberous sclerosis (TSC1 and TSC2), neurofibromatosis (NF1), or Angelman syndrome (UBE3A). *FMR1* encodes the FMRP, which is linked to fragile X syndrome. Also, ASD-associated mutations have been found in neuroligin 3 and 4 (*NLGN3* and *NLGN4X*), neurexin 1 (*NRXN1*), the family of discs large (*DLG*) genes, for example, in *DLG4* (*PSD-95*), *ERK1* (*MAPK3*), and synaptic Ras GTPase-activating protein (*SynGAP1*). These proteins are all found in a complex with SHANK2 and SHANK3 at synapses, where they belong to two pathways that, if impaired, result in aberrant mRNA translation of synaptic proteins and abnormal synapse formation. For example, TSC1,

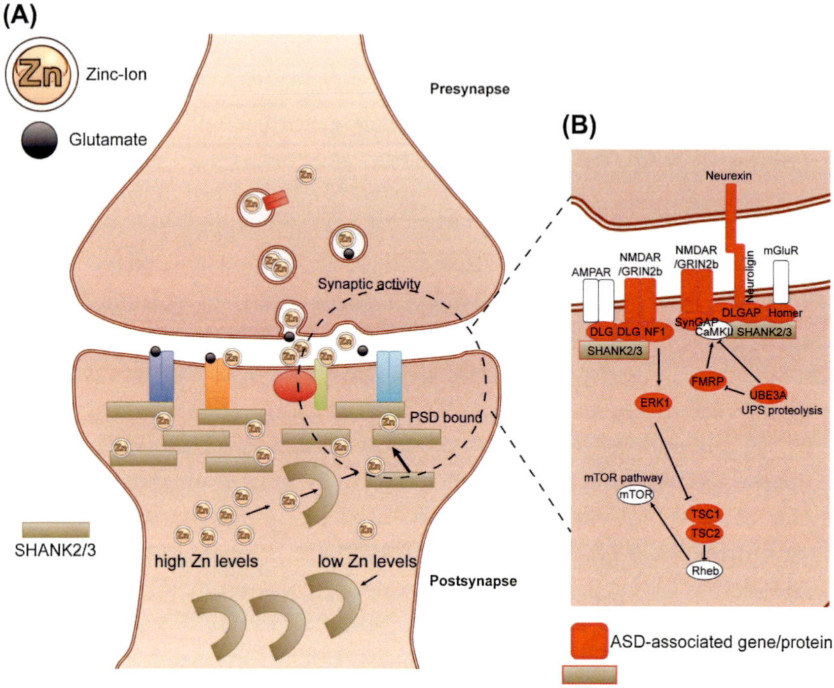

Figure 8.1 (A) A model for postsynaptic zinc signaling via SHANK2 and SHANK3 at glutamatergic synapses. At zincergic synapses, synaptic activity leads to the release of both glutamate and zinc into the synaptic cleft. There, zinc binds to AMPAR, NMDAR, or GPR39 receptors or enters into the postsynaptic compartment via channels such as NMDAR, AMPAR, and Ca_V. In addition, and possibly at the majority of all glutamatergic synapses, synaptic activity leads to the release of zinc from MT3 or other intracellular zinc stores. Postsynaptic zinc can bind SHANK2 and SHANK3 proteins that are part of a soluble pool and PSD-bound pool, shifting some molecules to the PSD-bound pool and thus stabilizing the PSD scaffold. However, inactivity and low postsynaptic zinc levels will favor the presence of SHANK2/3 in the soluble pool. (B) SHANK proteins are at the center of a signaling complex that is highly associated with ASD through mutations found in human patients (orange-labeled proteins). This complex regulates, among others, trafficking of AMPAR subunits to the postsynaptic membrane, levels of NMDAR, transsynaptic signaling via neuroligin–neurexin complexes, the degradation of proteins via the ubiquitin/proteasome system (UPS), and protein synthesis via the mTOR pathway and, thereby ultimately, regulates the organization of the PSD, cytoskeleton dynamics, cellular signaling cascades, epigenetic regulation of transcription, and release of neurotransmitters.

TSC2, and NF1 inhibit translation through the PI3K-mTOR signaling pathway, and FMRP inhibits mRNA translation at the synapse. NLGN, NRXN, DLG, SynGAP, and SHANK are important mediators of synaptogenesis and synaptic plasticity.

The loss of SHANK protein function is tightly linked to pathologies of the CNS, together termed Shankopathies (SHANK protein deficiency–induced synaptopathies) (Ey et al., 2020; Grabrucker et al., 2011b). Shankopathies include ASD, Phelan McDermid syndrome, intellectual disability, schizophrenia, but also other disorders where abnormal SHANK levels and function have also been observed (e.g., Alzheimer's disease (Grabrucker et al., 2011c) and bipolar disorder (Zhang et al., 2016)).

Mice developing under zinc deficiency in utero not only display ASD-like behavior (see Chapter 9) but also show lasting alterations in the composition of glutamatergic synapses. Indeed, low zinc levels during development lead to a significant reduction in SHANK proteins at the PSD of synapses, as well as the number of subunits of NMDA and AMPA receptors (Grabrucker et al., 2014). The data show that zinc levels during pregnancy are critical for the regulation of ASD-associated synaptic proteins in vivo. Thus, zinc deficiency as an environmental factor for ASD is linked with the synaptic ASD-associated pathway identified from candidate-gene studies via regulation of SHANK proteins (Grabrucker, 2014).

Interestingly, although zinc levels were restored after birth and older animals had normal zinc levels, the transient loss of SHANK proteins during a critical window in brain development in utero led to lasting changes in behavior, which reveals parallels to the measurement of trace metal imbalances, and in particular zinc deficiency, in humans that occurred predominantly early in life, despite lasting ASD symptoms.

Zinc and other autism spectrum disorder candidate genes

Data from mouse models show that prenatal zinc deficiency results in more noticeable effects at excitatory than inhibitory synapses, possibly due to the physiological role of zinc at zincergic synapses and the presence of SHANK proteins that are mostly limited to excitatory synapses. Apart from SHANK proteins, zinc binds to several receptors at the excitatory postsynaptic side that have been linked to ASD as well (Sceniak et al., 2019). For example, zinc inhibits NMDAR function by binding to the NR2A subunit (Paoletti et al., 1997; Rachline et al., 2005) and modulates the transmission of AMPAR (Blakemore and Trombley, 2017). However, reports have also shown that zinc can affect inhibitory receptors such as γ-aminobutyric acid (GABA) and glycine amino acid receptors as well (Nguyen et al., 2019). Together, the effects of zinc deficiency may lead to an imbalance between excitation and inhibition that has long been suggested as an underlying factor in ASD (Canitano and Pallagrosi, 2017).

Several other ASD candidate genes play a role in zinc homeostasis and signaling. For example, the metal regulatory transcription factor 1 (*MTF1*), a zinc-binding transcription factor that regulates the expression of metallothionein genes such as MT3, has been associated with ASD (Serajee et al., 2004). MT3 plays a crucial role in releasing zinc for the regulation of SHANK3 (Grabrucker et al., 2014). Moreover, mutations in the neurotrophic receptor tyrosine kinase 2 (*NTRK2*) gene, encoding for the TrkB receptor, have been associated with developmental delay and ASD (Hamdan et al., 2017). Zinc activates the TrkB receptor that is localized at excitatory synapses in an activity-dependent mechanism (Huang et al., 2008). The signaling downstream of TrkB, but also NMDA receptors, is mediated by the activation of the Mitogen-activated protein kinase 3 (*MAPK3, ERK1*). Mutations in the *MAPK3* gene have been linked to ASD (Schaaf et al., 2011). It has been demonstrated that zinc is an important modulator of the MAPK3 signaling pathway. MAPK signaling is critical for synaptic plasticity. Zinc deficiency results in increased phosphorylation of MAPK3 in vitro and in vivo. The abnormal MAPK3 phosphorylation is associated with an altered neural progenitor cell proliferation and cell death in the brain (Nuttall and Oteiza, 2012).

Furthermore, mutations in the zinc transporter ZnT5 (*SLC39A5*) have been reported in ASD (Gregg et al., 2008; O'Roak et al., 2011; Sanders et al., 2012). ZnT5 expression has been reported in the brain. However, its exact role remains unknown so far. ZnT5 is also expressed in the gastrointestinal system and the placenta (Ford, 2004). Therefore, loss of function may impair prenatal placental zinc transport and thereby contribute to abnormal brain zinc homeostasis, affecting brain development and functions.

Several more ASD candidate genes encode for zinc-binding proteins. Among these are *DPP3, LMX1B, SOD1, RNF38,* and *AZGP1*. DDP3 and SOD1 play a role in protection against oxidative stress. LMX1B is a transcription factor, and RNF38 is an E3 ubiquitin protein ligase, which is similar to UBE3A, an essential protein in the etiology of Angelman Syndrome (Buiting et al., 2016). The role of AZGP1 in the CNS is less clear. However, the exact number of zinc-binding proteins in the Simons Foundation Autism Research Initiative (SFARI) Gene database has not been investigated so far. Nevertheless, in addition to the genes mentioned earlier, the database currently lists further 31 ASD candidate genes with a zinc-binding zinc finger protein domain (Fig. 8.2). A Gene Ontology enrichment analysis (PANTHER Overrepresentation Test) investigating the biological process that the zinc-binding ASD candidate genes (Fig. 8.2)

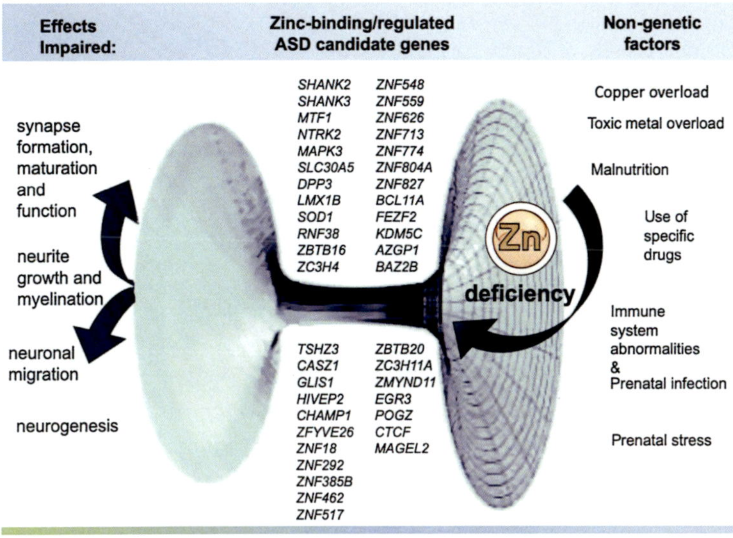

Figure 8.2 Several nongenetic and genetic risk factors for ASD impact on zinc homeostasis. In the developing brain, altered zinc availability will affect the function of zinc-binding proteins. Indeed, several key ASD candidate genes such as *SHANK2* and *SHANK3* encode for zinc-binding proteins that are involved in signaling pathways that ultimately regulate synapse formation, maturation, and function (LTP, E/I balance, and brain connectivity), neurite outgrowth, neuronal migration, and neurogenesis. The list of genes is taken from the SFARI gene bank. However, several more ASD candidate genes may bind zinc or may be regulated by zinc. *ASD*, autism spectrum disorder.

are involved in results in a significant ($P < .05$) association with (1) synaptic growth at the neuromuscular junction (nervous system development), (2) positive regulation of synaptic transmission, glutamatergic, (3) long-term synaptic potentiation (LTP), and (4) negative and positive regulation of transcription, DNA-templated.

Zinc deficiency and its relation to other autism spectrum disorder risk factors

Apart from the proteins listed earlier that bind zinc and thus may be susceptible to zinc deficiency, several known nongenetic risk factors for ASD affect zinc homeostasis (Grabrucker, 2012) (Fig. 8.2).

For example, zinc deficiency affects the immune system. On the organism level, this is evident through the association of a low zinc status with an increased susceptibility for infections (Read et al., 2019). In particular, zinc deficiency was shown to promote systemic inflammation, and a role for zinc for the growth and function of B and T cells, neutrophils, and natural killer cells and macrophages was reported (Wessels et al., 2017). However, the immune system also affects zinc homeostasis. During an infection, plasma zinc levels are significantly reduced. This is mediated through the release of S100 proteins and IL-6 that increases the expression of *ZIP14* in hepatocytes. This, in turn, results in uptake of zinc into the liver, thereby reducing serum zinc levels (Gammoh and Rink, 2017) that are the source of zinc for the developing embryo. Similarly, prenatal stress may also influence the immune system, and by that, zinc levels (Veru et al., 2014).

Apart from maternal malnutrition, a low zinc status may be caused by the exposure to excessive toxic metals and copper overload since these metals may compete with zinc ions for protein binding. Besides, the use of specific drugs during pregnancy, such as valproate (valproic acid, sodium valproate) that is used in the treatment of epilepsy, bipolar disorder, and migraine, has been linked to ASD. Prenatal exposure to valproate increases the risk for ASD (Christensen et al., 2013) in humans and induces ASD-like behavior in rodents (Nicolini and Fahnestock, 2018). Some side effects of valproate treatment are similar to the effects of zinc deficiency, and it has been hypothesized that valproate affects zinc homeostasis. Interestingly, some behavioral impairments of the valproic acid–induced rodent model of ASD can be rescued by zinc treatment (Cezar et al., 2018).

Conclusions

There is strong evidence that zinc plays a critical role in key processes that, if impaired, provide the biological pathomechanism for ASD. It is likely that zinc deficiency will not affect all zinc-binding proteins and thus all zinc-dependent processes in the brain equally. The effects of zinc deficiency will be subject to the severity of the deficiency and the affinity of the zinc-binding site of the involved proteins for zinc. High-affinity zinc-binding sites that are often found in transcription factors (zinc finger domains) may only be affected in case a severe zinc deficiency has occurred. This may be the reason why only severe, but not mild, zinc deficiency during pregnancy shows teratological effects (da Cunha Ferreira et al., 1989). However, synaptic processes such as the dynamic regulation of SHANK2 and SHANK3, NMDAR, or MAPK signaling are based on free zinc or weakly bound zinc ions. These processes will already be affected by low levels of zinc depletion. In mouse models, induction of a mild zinc deficiency during pregnancy was sufficient to induce ASD-like behaviors (Grabrucker et al., 2014, 2016). In line with this, the proteins that are part of zinc signaling at synapses and that are among those that utilize free zinc or weakly bound zinc are the major ASD candidate genes that are also frequently linked to syndromic forms of ASD.

Together, the current data make prenatal zinc deficiency a strong candidate for a causative factor for ASD and may explain the large share of individuals with ASD, where no mutation in ASD candidate genes can be detected. Given that many encoded proteins of ASD candidate genes are part of zinc-regulated processes, the involved signaling pathways such as the NRXN-NLGN-SHANK or the mTOR/PI3K pathway are zinc-dependent. While no mutation of the proteins may have occurred, zinc deficiency, which may no longer be detectable in the individual with ASD at the time of diagnosis, could have impaired the pathways similar to mutations in their key proteins. The fact that prenatal zinc-deficient animals show ASD-like behavior (see Chapter 9) together with detectable alterations in the ASD-associated signaling pathways makes zinc the trace metal whose relationship to ASD has been explained best on a mechanistic level. Intriguingly, several other environmental risk factors for ASD affect zinc homeostasis and, therefore, may converge on a shared biological mechanism that is centered around a low zinc availability for critical neuronal processes.

References

Arons, M.H., Thynne, C.J., Grabrucker, A.M., Li, D., Schoen, M., Cheyne, J.E., Boeckers, T.M., Montgomery, J.M., Garner, C.C., 2012. Autism-associated mutations in ProSAP2/SHANK3 impair synaptic transmission and neurexin-neuroligin-mediated transsynaptic signaling. J Neurosci 32 (43), 14966–14978.

Baron, M.K., Boeckers, T.M., Vaida, B., Faham, S., Gingery, M., Sawaya, M.R., Salyer, D., Gundelfinger, E.D., Bowie, J.U., 2006. An architectural framework that may lie at the core of the postsynaptic density. Science 311 (5760), 531–535.

Berkel, S., Marshall, C.R., Weiss, B., Howe, J., Roeth, R., Moog, U., Endris, V., Roberts, W., Szatmari, P., Pinto, D., Bonin, M., Riess, A., Engels, H., Sprengel, R., Scherer, S.W., Rappold, G.A., 2010. Mutations in the SHANK2 synaptic scaffolding gene in autism spectrum disorder and mental retardation. Nat Genet 42 (6), 489–491.

Blakemore, L.J., Trombley, P.Q., 2017. Zinc as a neuromodulator in the central nervous system with a focus on the olfactory bulb. Front Cell Neurosci 11, 297.

Boeckers, T.M., 2006. The postsynaptic density. Cell Tissue Res 326 (2), 409–422.

Boeckers, T.M., Bockmann, J., Kreutz, M.R., Gundelfinger, E.D., 2002. ProSAP/Shank proteins – a family of higher order organizing molecules of the postsynaptic density with an emerging role in human neurological disease. J Neurochem 81 (5), 903–910.

Boeckers, T.M., Liedtke, T., Spilker, C., Dresbach, T., Bockmann, J., Kreutz, M.R., Gundelfinger, E.D., 2005. C-terminal synaptic targeting elements for postsynaptic density proteins ProSAP1/SHANK2 and ProSAP2/SHANK3. J Neurochem 92 (3), 519–524.

Bourgeron, T., 2009. A synaptic trek to autism. Curr Opin Neurobiol 19 (2), 231–234.

Bourgeron, T., 2015. From the genetic architecture to synaptic plasticity in autism spectrum disorder. Nat Rev Neurosci 16 (9), 551–563.

Buiting, K., Williams, C., Horsthemke, B., 2016. Angelman syndrome – insights into a rare neurogenetic disorder. Nat Rev Neurol 12 (10), 584–593.

Canitano, R., Pallagrosi, M., 2017. Autism spectrum disorders and schizophrenia spectrum disorders: excitation/inhibition imbalance and developmental trajectories. Front Psychiatr 8, 69.

Cezar, L.C., Kirsten, T.B., da Fonseca, C.C.N., de Lima, A.P.N., Bernardi, M.M., Felicio, L.F., 2018. Zinc as a therapy in a rat model of autism prenatally induced by valproic acid. Prog Neuro-Psychopharmacol Biol Psychiatry 84 (Pt A), 173–180.

Chen, J., Yu, S., Fu, Y., Li, X., 2014. Synaptic proteins and receptors defects in autism spectrum disorders. Front Cell Neurosci 8, 276.

Christensen, J., Grønborg, T.K., Sørensen, M.J., Schendel, D., Parner, E.T., Pedersen, L.H., Vestergaard, M., 2013. Prenatal valproate exposure and risk of autism spectrum disorders and childhood autism. J Am Med Assoc 309 (16), 1696–1703.

Costales, J.L., Kolevzon, A., 2015. Phelan-McDermid syndrome and SHANK3: implications for treatment. Neurotherapeutics 12 (3), 620–630.

da Cunha Ferreira, R.M., Marquiegui, I.M., Elizaga, I.V., 1989. Teratogenicity of zinc deficiency in the rat: study of the fetal skeleton. Teratology 39 (2), 181–194.

Durand, C.M., Betancur, C., Boeckers, T.M., Bockmann, J., Chaste, P., Fauchereau, F., Nygren, G., Rastam, M., Gillberg, I.C., Anckarsäter, H., Sponheim, E., Goubran-Botros, H., Delorme, R., Chabane, N., Mouren-Simeoni, M.C., de Mas, P., Bieth, E., Rogé, B., Héron, D., Burglen, L., Gillberg, C., Leboyer, M., Bourgeron, T., 2007. Mutations in the gene encoding the synaptic scaffolding protein SHANK3 are associated with autism spectrum disorders. Nat Genet 39 (1), 25–27.

Ey, E., Bourgeron, T., Boeckers, T.M., Kim, E., Han, K., 2020. Editorial: Shankopathies: shank protein deficiency-induced synaptic diseases. Front Mol Neurosci 13, 11.

Ford, D., 2004. Intestinal and placental zinc transport pathways. Proc Nutr Soc 63 (1), 21–29.

Gammoh, N.Z., Rink, L., 2017. Zinc in infection and inflammation. Nutrients 9 (6).

Gauthier, J., Spiegelman, D., Piton, A., Lafrenière, R.G., Laurent, S., St-Onge, J., Lapointe, L., Hamdan, F.F., Cossette, P., Mottron, L., Fombonne, E., Joober, R., Marineau, C., Drapeau, P., Rouleau, G.A., 2009. Novel de novo SHANK3 mutation in autistic patients. Am J Med Genet B Neuropsychiatr Genet 150B (3), 421–424.

Gilbert, J., Man, H.Y., 2017. Fundamental elements in autism: from neurogenesis and neurite growth to synaptic plasticity. Front Cell Neurosci 11, 359.

Gonzalez-Gronow, M., Cuchacovich, M., Francos, R., Cuchacovich, S., Blanco, A., Sandoval, R., Gomez, C.F., Valenzuela, J.A., Ray, R., Pizzo, S.V., 2015. Catalytic autoantibodies against myelin basic protein (MBP) isolated from serum of autistic children impair in vitro models of synaptic plasticity in rat hippocampus. J Neuroimmunol 287, 1–8.

Grabrucker, A.M., 2012. Environmental factors in autism. Front Psychiatr 3, 118.

Grabrucker, A.M., 2014. A role for synaptic zinc in ProSAP/Shank PSD scaffold malformation in autism spectrum disorders. Dev Neurobiol 74 (2), 136–146.

Grabrucker, A.M., Knight, M.J., Proepper, C., Bockmann, J., Joubert, M., Rowan, M., Nienhaus, G.U., Garner, C.C., Bowie, J.U., Kreutz, M.R., Gundelfinger, E.D.,

Boeckers, T.M., 2011a. Concerted action of zinc and ProSAP/Shank in synaptogenesis and synapse maturation. EMBO J 30 (3), 569–581.
Grabrucker, A.M., Schmeisser, M.J., Schoen, M., Boeckers, T.M., 2011b. Postsynaptic ProSAP/Shank scaffolds in the cross-hair of synaptopathies. Trends Cell Biol 21 (10), 594–603.
Grabrucker, A.M., Schmeisser, M.J., Udvardi, P.T., Arons, M., Schoen, M., Woodling, N.S., Andreasson, K.I., Hof, P.R., Buxbaum, J.D., Garner, C.C., Boeckers, T.M., 2011c. Amyloid beta protein-induced zinc sequestration leads to synaptic loss via dysregulation of the ProSAP2/SHANK3 scaffold. Mol Neurodegener 6, 65.
Grabrucker, A.M., Vaida, B., Bockmann, J., Boeckers, T.M., 2009. Efficient targeting of proteins to post-synaptic densities of excitatory synapses using a novel pSDTarget vector system. J Neurosci Methods 181 (2), 227–234.
Grabrucker, S., Boeckers, T.M., Grabrucker, A.M., 2016. Gender dependent evaluation of autism like behavior in mice exposed to prenatal zinc deficiency. Front Behav Neurosci 10, 37.
Grabrucker, S., Jannetti, L., Eckert, M., Gaub, S., Chhabra, R., Pfaender, S., Mangus, K., Reddy, P.P., Rankovic, V., Schmeisser, M.J., Kreutz, M.R., Ehret, G., Boeckers, T.M., Grabrucker, A.M., 2014. Zinc deficiency dysregulates the synaptic ProSAP/Shank scaffold and might contribute to autism spectrum disorders. Brain 137 (Pt 1), 137–152.
Gregg, J.P., Lit, L., Baron, C.A., Hertz-Picciotto, I., Walker, W., Davis, R.A., Croen, L.A., Ozonoff, S., Hansen, R., Pessah, I.N., Sharp, F.R., 2008. Gene expression changes in children with autism. Genomics 91 (1), 22–29.
Guang, S., Pang, N., Deng, X., Yang, L., He, F., Wu, L., Chen, C., Yin, F., Peng, J., 2018. Synaptopathology involved in autism spectrum disorder. Front Cell Neurosci 12, 470.
Gundelfinger, E.D., Boeckers, T.M., Baron, M.K., Bowie, J.U., 2006. A role for zinc in postsynaptic density asSAMbly and plasticity? Trends Biochem Sci 31 (7), 366–373.
Ha, H.T.T., Leal-Ortiz, S., Lalwani, K., Kiyonaka, S., Hamachi, I., Mysore, S.P., Montgomery, J.M., Garner, C.C., Huguenard, J.R., Kim, S.A., 2018. Shank and zinc mediate an AMPA receptor subunit switch in developing neurons. Front Mol Neurosci 11, 405.
Hamdan, F.F., Myers, C.T., Cossette, P., Lemay, P., Spiegelman, D., Laporte, A.D., Nassif, C., Diallo, O., Monlong, J., Cadieux-Dion, M., Dobrzeniecka, S., Meloche, C., Retterer, K., Cho, M.T., Rosenfeld, J.A., Bi, W., Massicotte, C., Miguet, M., Brunga, L., Regan, B.M., Mo, K., Tam, C., Schneider, A., Hollingsworth, G., FitzPatrick, D.R., Donaldson, A., Canham, N., Blair, E., Kerr, B., Fry, A.E., Thomas, R.H., Shelagh, J., Hurst, J.A., Brittain, H., Blyth, M., Lebel, R.R., Gerkes, E.H., Davis-Keppen, L., Stein, Q., Chung, W.K., Dorison, S.J., Benke, P.J., Fassi, E., Corsten-Janssen, N., Kamsteeg, E.J., Mau-Them, F.T., Bruel, A.L., Verloes, A., Õunap, K., Wojcik, M.H., Albert, D.V.F., Venkateswaran, S., Ware, T., Jones, D., Liu, Y.C., Mohammad, S.S., Bizargity, P., Bacino, C.A., Leuzzi, V., Martinelli, S., Dallapiccola, B., Tartaglia, M., Blumkin, L., Wierenga, K.J., Purcarin, G., O'Byrne, J.J., Stockler, S., Lehman, A., Keren, B., Nougues, M.C., Mignot, C., Auvin, S., Nava, C., Hiatt, S.M., Bebin, M., Shao, Y., Scaglia, F., Lalani, S.R., Frye, R.E., Jarjour, I.T., Jacques, S., Boucher, R.M., Riou, E., Srour, M., Carmant, L., Lortie, A., Major, P., Diadori, P., Dubeau, F., D'Anjou, G., Bourque, G., Berkovic, S.F., Sadleir, L.G., Campeau, P.M., Kibar, Z., Lafrenière, R.G., Girard, S.L., Mercimek-Mahmutoglu, S., Boelman, C., Rouleau, G.A., Scheffer, I.E., Mefford, H.C., Andrade, D.M., Rossignol, E., Minassian, B.A., Michaud, J.L., 2017. High rate of recurrent de novo mutations in developmental and epileptic encephalopathies. Am J Hum Genet 101 (5), 664–685.

Huang, Y.Z., Pan, E., Xiong, Z.Q., McNamara, J.O., 2008. Zinc-mediated transactivation of TrkB potentiates the hippocampal mossy fiber-CA3 pyramid synapse. Neuron 57 (4), 546–558.
Huguet, G., Ey, E., Bourgeron, T., 2013. The genetic landscapes of autism spectrum disorders. Annu Rev Genom Hum Genet 14, 191–213.
Jan, H.H., Chen, I.T., Tsai, Y.Y., Chang, Y.C., 2002. Structural role of zinc ions bound to postsynaptic densities. J Neurochem 83 (3), 525–534.
Leblond, C.S., Nava, C., Polge, A., Gauthier, J., Huguet, G., Lumbroso, S., Giuliano, F., Stordeur, C., Depienne, C., Mouzat, K., Pinto, D., Howe, J., Lemière, N., Durand, C.M., Guibert, J., Ey, E., Toro, R., Peyre, H., Mathieu, A., Amsellem, F., Rastam, M., Gillberg, I.C., Rappold, G.A., Holt, R., Monaco, A.P., Maestrini, E., Galan, P., Heron, D., Jacquette, A., Afenjar, A., Rastetter, A., Brice, A., Devillard, F., Assouline, B., Laffargue, F., Lespinasse, J., Chiesa, J., Rivier, F., Bonneau, D., Regnault, B., Zelenika, D., Delepine, M., Lathrop, M., Sanlaville, D., Schluth-Bolard, C., Edery, P., Perrin, L., Tabet, A.C., Schmeisser, M.J., Boeckers, T.M., Coleman, M., Sato, D., Szatmari, P., Scherer, S.W., Rouleau, G.A., Betancur, C., Leboyer, M., Gillberg, C., Delorme, R., Bourgeron, T., 2014. Meta-analysis of SHANK mutations in autism spectrum disorders: a gradient of severity in cognitive impairments. PLoS Genet 10 (9), e1004580.
LiCausi, F., Hartman, N.W., 2018. Role of mTOR complexes in neurogenesis. Int J Mol Sci 19 (5).
Meyer, G., Varoqueaux, F., Neeb, A., Oschlies, M., Brose, N., 2004. The complexity of PDZ domain-mediated interactions at glutamatergic synapses: a case study on neuroligin. Neuropharmacology 47 (5), 724–733.
Moessner, R., Marshall, C.R., Sutcliffe, J.S., Skaug, J., Pinto, D., Vincent, J., Zwaigenbaum, L., Fernandez, B., Roberts, W., Szatmari, P., Scherer, S.W., 2007. Contribution of SHANK3 mutations to autism spectrum disorder. Am J Hum Genet 81 (6), 1289–1297.
Naisbitt, S., Kim, E., Tu, J.C., Xiao, B., Sala, C., Valtschanoff, J., Weinberg, R.J., Worley, P.F., Sheng, M., 1999. Shank, a novel family of postsynaptic density proteins that binds to the NMDA receptor/PSD-95/GKAP complex and cortactin. Neuron 23 (3), 569–582.
Nicolini, C., Fahnestock, M., 2018. The valproic acid-induced rodent model of autism. Exp Neurol 299 (Pt A), 217–227.
Nguyen, H.T., Jang, S.H., Park, S.J., Cho, D.H., Han, S.K., 2019. Modulation of inhibitory and excitatory neurotransmissions by Zn2+ on the substantia gelatinosa neurons of the trigeminal subnucleus caudalis in mice. Gen Physiol Biophys 38 (5), 399–406.
Nuttall, J.R., Oteiza, P.I., 2012. Zinc and the ERK kinases in the developing brain. Neurotox Res 21 (1), 128–141.
Oberman, L.M., Boccuto, L., Cascio, L., Sarasua, S., Kaufmann, W.E., 2015. Autism spectrum disorder in Phelan-McDermid syndrome: initial characterization and genotype-phenotype correlations. Orphanet J Rare Dis 10, 105.
O'Roak, B.J., Deriziotis, P., Lee, C., Vives, L., Schwartz, J.J., Girirajan, S., Karakoc, E., Mackenzie, A.P., Ng, S.B., Baker, C., Rieder, M.J., Nickerson, D.A., Bernier, R., Fisher, S.E., Shendure, J., Eichler, E.E., 2011. Exome sequencing in sporadic autism spectrum disorders identifies severe de novo mutations. Nat Genet 43 (6), 585–589.
Paoletti, P., Ascher, P., Neyton, J., 1997. High-affinity zinc inhibition of NMDA NR1-NR2A receptors. J Neurosci 17 (15), 5711–5725.
Pinto, D., Pagnamenta, A.T., Klei, L., Anney, R., Merico, D., Regan, R., Conroy, J., Magalhaes, T.R., Correia, C., Abrahams, B.S., Almeida, J., Bacchelli, E., Bader, G.D., Bailey, A.J., Baird, G., Battaglia, A., Berney, T., Bolshakova, N., Bölte, S., Bolton, P.F., Bourgeron, T., Brennan, S., Brian, J., Bryson, S.E., Carson, A.R., Casallo, G., Casey, J.,

Chung, B.H., Cochrane, L., Corsello, C., Crawford, E.L., Crossett, A., Cytrynbaum, C., Dawson, G., de Jonge, M., Delorme, R., Drmic, I., Duketis, E., Duque, F., Estes, A., Farrar, P., Fernandez, B.A., Folstein, S.E., Fombonne, E., Freitag, C.M., Gilbert, J., Gillberg, C., Glessner, J.T., Goldberg, J., Green, A., Green, J., Guter, S.J., Hakonarson, H., Heron, E.A., Hill, M., Holt, R., Howe, J.L., Hughes, G., Hus, V., Igliozzi, R., Kim, C., Klauck, S.M., Kolevzon, A., Korvatska, O., Kustanovich, V., Lajonchere, C.M., Lamb, J.A., Laskawiec, M., Leboyer, M., Le Couteur, A., Leventhal, B.L., Lionel, A.C., Liu, X.Q., Lord, C., Lotspeich, L., Lund, S.C., Maestrini, E., Mahoney, W., Mantoulan, C., Marshall, C.R., McConachie, H., McDougle, C.J., McGrath, J., McMahon, W.M., Merikangas, A., Migita, O., Minshew, N.J., Mirza, G.K., Munson, J., Nelson, S.F., Noakes, C., Noor, A., Nygren, G., Oliveira, G., Papanikolaou, K., Parr, J.R., Parrini, B., Paton, T., Pickles, A., Pilorge, M., Piven, J., Ponting, C.P., Posey, D.J., Poustka, A., Poustka, F., Prasad, A., Ragoussis, J., Renshaw, K., Rickaby, J., Roberts, W., Roeder, K., Roge, B., Rutter, M.L., Bierut, L.J., Rice, J.P., Salt, J., Sansom, K., Sato, D., Segurado, R., Sequeira, A.F., Senman, L., Shah, N., Sheffield, V.C., Soorya, L., Sousa, I., Stein, O., Sykes, N., Stoppioni, V., Strawbridge, C., Tancredi, R., Tansey, K., Thiruvahindrapduram, B., Thompson, A.P., Thomson, S., Tryfon, A., Tsiantis, J., Van Engeland, H., Vincent, J.B., Volkmar, F., Wallace, S., Wang, K., Wang, Z., Wassink, T.H., Webber, C., Weksberg, R., Wing, K., Wittemeyer, K., Wood, S., Wu, J., Yaspan, B.L., Zurawiecki, D., Zwaigenbaum, L., Buxbaum, J.D., Cantor, R.M., Cook, E.H., Coon, H., Cuccaro, M.L., Devlin, B., Ennis, S., Gallagher, L., Geschwind, D.H., Gill, M., Haines, J.L., Hallmayer, J., Miller, J., Monaco, A.P., Nurnberger, J.I., Paterson, A.D., Pericak-Vance, M.A., Schellenberg, G.D., Szatmari, P., Vicente, A.M., Vieland, V.J., Wijsman, E.M., Scherer, S.W., Sutcliffe, J.S., Betancur, C., 2010. Functional impact of global rare copy number variation in autism spectrum disorders. Nature 466 (7304), 368–372.

Rachline, J., Perin-Dureau, F., Le Goff, A., Neyton, J., Paoletti, P., 2005. The micromolar zinc-binding domain on the NMDA receptor subunit NR2B. J Neurosci 25 (2), 308–317.

Read, S.A., Obeid, S., Ahlenstiel, C., Ahlenstiel, G., 2019. The role of zinc in antiviral immunity. Adv Nutr 10 (4), 696–710.

Sanders, S.J., Murtha, M.T., Gupta, A.R., Murdoch, J.D., Raubeson, M.J., Willsey, A.J., Ercan-Sencicek, A.G., DiLullo, N.M., Parikshak, N.N., Stein, J.L., Walker, M.F., Ober, G.T., Teran, N.A., Song, Y., El-Fishawy, P., Murtha, R.C., Choi, M., Overton, J.D., Bjornson, R.D., Carriero, N.J., Meyer, K.A., Bilguvar, K., Mane, S.M., Sestan, N., Lifton, R.P., Günel, M., Roeder, K., Geschwind, D.H., Devlin, B., State, M.W., 2012. De novo mutations revealed by whole-exome sequencing are strongly associated with autism. Nature 485 (7397), 237–241.

Sanford, L., Carpenter, M.C., Palmer, A.E., 2019. Intracellular Zn^{2+} transients modulate global gene expression in dissociated rat hippocampal neurons. Sci Rep 9 (1), 9411.

Sceniak, M.P., Fedder, K.N., Wang, Q., Droubi, S., Babcock, K., Patwardhan, S., Wright-Zornes, J., Pham, L., Sabo, S.L., 2019. An autism-associated mutation in GluN2B prevents NMDA receptor trafficking and interferes with dendrite growth. J Cell Sci 132 (20).

Schaaf, C.P., Sabo, A., Sakai, Y., Crosby, J., Muzny, D., Hawes, A., Lewis, L., Akbar, H., Varghese, R., Boerwinkle, E., Gibbs, R.A., Zoghbi, H.Y., 2011. Oligogenic heterozygosity in individuals with high-functioning autism spectrum disorders. Hum Mol Genet 20 (17), 3366–3375.

Schmeisser, M.J., Ey, E., Wegener, S., Bockmann, J., Stempel, A.V., Kuebler, A., Janssen, A.L., Udvardi, P.T., Shiban, E., Spilker, C., Balschun, D., Skryabin, B.V., Dieck, S., Smalla, K.H., Montag, D., Leblond, C.S., Faure, P., Torquet, N., Le

Sourd, A.M., Toro, R., Grabrucker, A.M., Shoichet, S.A., Schmitz, D., Kreutz, M.R., Bourgeron, T., Gundelfinger, E.D., Boeckers, T.M., 2012. Autistic-like behaviours and hyperactivity in mice lacking ProSAP1/SHANK2. Nature 486 (7402), 256–260.

Serajee, F.J., Nabi, R., Zhong, H., Huq, M., 2004. Polymorphisms in xenobiotic metabolism genes and autism. J Child Neurol 19 (6), 413–417.

Tao-Cheng, J.H., Toy, D., Winters, C.A., Reese, T.S., Dosemeci, A., 2016. Zinc stabilizes Shank3 at the postsynaptic density of hippocampal synapses. PLoS One 11 (5), e0153979.

Uchino, S., Wada, H., Honda, S., Nakamura, Y., Ondo, Y., Uchiyama, T., Tsutsumi, M., Suzuki, E., Hirasawa, T., Kohsaka, S., 2006. Direct interaction of post-synaptic density-95/Dlg/ZO-1 domain-containing synaptic molecule Shank3 with GluR1 alpha-amino-3-hydroxy-5-methyl-4-isoxazole propionic acid receptor. J Neurochem 97 (4), 1203–1214.

Verpelli, C., Dvoretskova, E., Vicidomini, C., Rossi, F., Chiappalone, M., Schoen, M., Di Stefano, B., Mantegazza, R., Broccoli, V., Böckers, T.M., Dityatev, A., Sala, C., 2011. Importance of Shank3 protein in regulating metabotropic glutamate receptor 5 (mGluR5) expression and signaling at synapses. J Biol Chem 286 (40), 34839–34850.

Veru, F., Laplante, D.P., Luheshi, G., King, S., 2014. Prenatal maternal stress exposure and immune function in the offspring. Stress 17 (2), 133–148.

Wessels, I., Maywald, M., Rink, L., 2017. Zinc as a gatekeeper of immune function. Nutrients 9 (12).

Won, H., Mah, W., Kim, E., 2013. Autism spectrum disorder causes, mechanisms, and treatments: focus on neuronal synapses. Front Mol Neurosci 6, 19.

Yang, P., Chang, C.L., 2014. Glutamate-mediated signaling and autism spectrum disorders: emerging treatment targets. Curr Pharmaceut Des 20 (32), 5186–5193.

Zhang, C., Wu, Z., Hong, W., Peng, D., Fang, Y., 2016. Evaluating the association between the SHANK3 gene and bipolar disorder. Psychiatr Res 244, 284–288.

Zoghbi, H.Y., Bear, M.F., 2012. Synaptic dysfunction in neurodevelopmental disorders associated with autism and intellectual disabilities. Cold Spring Harb Perspect Biol 4 (3).

CHAPTER 9

Animal models for trace metal abnormalities—links to autism

Contents

Introduction

Animal models are necessary to understand the impact of trace metal abnormalities on the whole organism. In particular, as there is no biomarker for ASD, in humans as well as animal models, an investigation of behavior is necessary. Thus, animals that communicate and display social behavior are critical to study biometals in ASD. In recent years, mouse models have been in the focus of ASD research, where overload and deficiencies in specific metals can be induced. To investigate ASD-like behavior in mice, a key set of assays has been proposed (Crawley, 2007). These behavioral tests are designed to evaluate the core features of ASD behavior (abnormal social interactions, deficits in communication, and high levels of repetitive behaviors), as well as the most prominent comorbidities such as cognitive impairments, hyperactivity, and anxiety.

Abnormal social behavior, for example, can be evaluated in a three-chambered sociability and social novelty test and by assessing reciprocal social interaction between two unfamiliar mice. Deficits in communication can be detected by analyzing (ultrasonic) vocalizations of mice. Repetitive behaviors can be evaluated by measuring the time animals spend on grooming, jumping, or sniffing, while a need for sameness can be detected by quantifying choices in tests for spatial memory (reversal of the spatial location of a platform in the Morris water maze or alternation of arm choices

Biometals in Autism Spectrum Disorders
ISBN 978-0-12-821132-8
https://doi.org/10.1016/B978-0-12-821132-8.00009-4

in the T-maze test). The latter also allows for gathering information on cognitive abilities. Further, marble burying can be used to assess repetitive behaviors (Crawley, 2007; Silverman et al., 2010). These tests are usually combined with an open field test that evaluates activity levels and anxiety, and an elevated plus-maze that also evaluates anxiety levels. Nest building is assessed to investigate cognitive impairments, although nest building may also be regarded as social behavior. Although not all ASD mouse models show abnormalities in all tests, reflecting the heterogeneity that can also be observed in human patients regarding their behavioral phenotypes, the overall picture emerging from these assays allows the classification of the behavior of the mouse as ASD-like. This may be followed by more specific assays that evaluate sleep patterns, responses to olfactory stimuli, memory, brain lateralization, or that assess the presence of seizures.

However, there is no standardized protocol for the induction of trace metal deficiencies and overload that has been used in studies investigating mouse models. Therefore the mice are treated with different concentrations of metals to induce overload, or they receive diets with different grades of trace metal depletion, for a different amount of time, and at different ages. Besides, gender effects have not been investigated in detail for some metals, where only male mice were used. Thus the severity and duration of metal deficiency and overload are different for many studies, which makes direct comparison difficult. Nevertheless, some overall trends can be observed for animals subjected to altered trace metal status. For ASD, the focus will be on the prenatal phase, where brain development occurs.

Zinc-deficient animals

As early as the 1970s, zinc-deficient animals have been generated and analyzed. Zinc deficiency can be induced by application of a zinc chelator, or by a zinc-depleted diet. Inducing zinc deficiency through diets may take 2–4 weeks until brain zinc levels are affected. In that case, zinc deficiency is an acute deficiency for the animal. Acute zinc-deficient animals may be pregnant and only zinc-deficient during the time of pregnancy. The pups of these mice often are raised under normal zinc conditions, and thus, serum zinc levels quickly recover, followed by tissue (brain) zinc levels. These animals are classified as prenatal zinc-deficient (PZD) animals. In addition, genetic knockout of zinc transporters and metallothioneins (MTs) may create systemic or local zinc deficiency and may affect different pools of zinc. For example, *ZnT3* (*Slc30a3*) knockout (KO) mice are devoid of vesicular zinc at synapses (Cole et al., 1999). However, the concentration of protein-bound zinc, even at synapses is not affected. *MT3* knockout mice, in contrast, have overall reduced brain zinc concentrations and impaired ability to buffer zinc and use weakly bound zinc to act as signaling ion in the brain. However, the presynaptic vesicular zinc seems unaffected (Erickson et al., 1997). Nevertheless, both *MT3* and *ZnT3* KO mice show some abnormal behavior that is related to ASD, hinting at the contribution of the depletion of both presynaptic and postsynaptic zinc stores to the ASD pathology. For example, *MT3* KO mice display diminished social interactions, mimicking aspects of ASDs (Koumura et al., 2009) and impaired synaptic transmission. *ZnT3* KO male mice show an increased cortical brain volume and also some ASD-like behaviors (Yoo et al., 2016).

The behavior of acute zinc-deficient and zinc transporter KO mice has been reviewed by Hagmeyer et al. (2014). Apart from *MT3* and *ZnT3* KO mice, only prenatal zinc-deficient mice show ASD-like behavior. However, most ZnT (*SLC30A*) and ZIP (*SLC39A*) KO mice have not been evaluated using behavioral phenotyping that is suitable to detect ASD (Kas et al., 2014) in mice so far.

Although there is some variability in the reports, the behavior of animal models for acute zinc deficiency can be summarized as being characterized by impaired learning, attention and memory, the presence of depression-like behavior, and increased aggression and anxiety (Hagmeyer et al., 2014). Especially, depression-like behavior seems to be a major phenotype (Tassabehji et al., 2008). However, acute zinc deficiency does not result in ASD-like behavior. In contrast, pre- and perinatal zinc deficiency is associated with altered learning, memory, anxiety, and ASD-like behavior and associated comorbidities (Fig. 9.1).

The core features of ASD are impaired communication and social behavior and repetitive behaviors. Already early studies investigating zinc deficiency in pregnant rhesus monkeys reported that their infants show altered social behavior (Sandstead et al., 1978). Altered social behavior of PZD rodents was also reported in multiple studies. For example, increased aggressiveness was observed in rats (Peters, 1978) and mice in a resident intruder test (Grabrucker et al., 2014). In addition, PZD mice display less maternal behavior toward their pups (Grabrucker et al., 2014) as well as impaired nesting behavior (Grabrucker et al., 2016). Besides, minor impairments were observed in the three-chamber box test (Grabrucker et al., 2016).

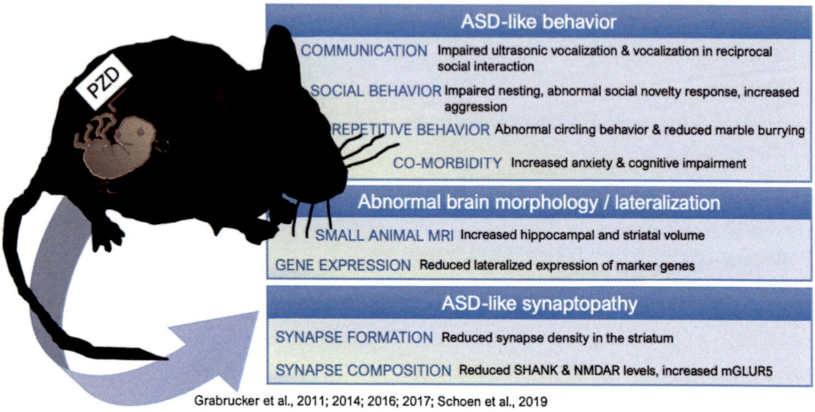

Figure 9.1 Zinc deficiency of pregnant mice during all weeks of pregnancy, and restoring zinc levels of pups after birth through nursing by foster mothers, leads to ASD-like behavior. PZD mice show abnormalities in tests assessing the three core features of ASD, communication and social impairments, and repetitive behaviors. Besides, PZD mice show increased anxiety and cognitive impairments. PZD induces specific changes in the gross morphology of selected brain regions and alters the expression of marker genes for brain lateralization. Besides, behavioral tests reveal abnormal brain lateralization of PZD mice. These effects may be caused by a loss of excitatory, but not inhibitory synapses in specific brain regions, and a synaptopathy that is characterized by dysregulation of proteins encoded by key ASD candidate genes.

Another core feature of ASD, communication deficits, was observed in PZD mice when they were adults. PZD mice show significant changes in ultrasonic vocalizations such as a decrease in the number of calls, sound pressure level, as well as the percentage of calls with overtones and harmonics (Grabrucker et al., 2016).

The third core feature, repetitive behaviors, is assessed by measuring time spent grooming, measuring circling behavior, and by assays such as the marble-burying test. The marble-burying behavior was abnormal in PZD female mice (Grabrucker et al., 2016). In addition, PZD mice show altered turning and circling behavior (Grabrucker et al., 2017) that have been associated with striatal defects (Grabrucker et al., 2017).

In addition to the ASD core features, comorbidities are frequent in ASD that comprise cognitive deficits, hyperactivity, increased anxiety, and epilepsy. Of these, cognitive impairments and increased anxiety are prominent in PZD mice. For example, mice with mild or moderate zinc deficiency from early gestation to around postnatal day 15 show learning and memory deficits as adults in a passive avoidance task (Golub et al., 1983). Similarly, pre- and perinatal zinc-deficient rats, when tested as adults, show memory problems in a radial maze test (Halas et al., 1983, 1986) and a Morris water maze test (Tahmasebi Boroujeni et al., 2009; Yu et al., 2013) as well as impaired motor learning in the rotarod assay (Grabrucker et al., 2016). In addition, the offspring of zinc-deficient rhesus monkeys have impaired learning later in life (Golub et al., 1985).

Enhanced stress response and emotionality after prenatal zinc deprivation were reported in various test paradigms (Lokken et al., 1973; Halas and Sandstead, 1975), and PZD mice display an anxiety phenotype in the open field test and the elevated plus-maze test (Grabrucker et al., 2016).

On a gross anatomical level, PZD mice show alterations in the brain structure of both the striatum and hippocampus (Grabrucker et al., 2017; Schoen et al., 2019) (Fig. 9.1). Some of the observed behavioral impairments may reflect these abnormalities. For example, the dysfunction of the striatum is associated with stereotypic and repetitive behaviors such as abnormal circling and altered marble burying that has been reported in PZD mice (Grabrucker et al., 2016, 2017). In addition, spatial learning impairment has been reported (Hagmeyer et al., 2014), a process that is dependent on a healthy hippocampus. Especially, the morphological differences in the hippocampus of PZD mice were maintained into adulthood (Schoen et al., 2019).

On the molecular level, the effects of zinc deficiency are plentiful due to the many zinc-binding and zinc-regulated proteins in the CNS. Not all of the many possibilities that may cause behavioral and structural abnormalities have been investigated in detail so far. However, among the reported effects, a clear connection to known molecular ASD pathologies emerged (Fig. 9.1). For example, PZD mice showed lasting alterations in the composition of excitatory glutamatergic synapses. A significant reduction of the ASD-associated postsynaptic density (PSD) proteins of the SHANK family (SHANK1, SHANK2, SHANK3), as well as subunits of NMDA and AMPA receptors (GluA1, GluN1, GluN2B), was observed, and an increase in synaptic mGluR5 proteins (Grabrucker et al., 2014). The reduction of SHANK proteins in PSD fractions was accompanied by an increase in the cytosolic fraction (Grabrucker et al., 2011, 2014) strengthening the model of a zinc-driven synaptic PSD integration of SHANK2 and SHANK3 proteins (Grabrucker, 2014; Tao-Cheng et al., 2016) (see Chapter 8). These findings are evidence that zinc deficiency affects the ASD-associated signaling pathway at synapses and link prenatal zinc deficiency mechanistically to known ASD candidate gene and their associated pathomechanisms.

Although some of the molecular changes are only transient and adult PZD mice, for example, do no longer show decreased synaptic SHANK2 and SHANK3 levels, reduced SHANK levels during a critical window in brain development and the associated impaired NRXN-NLGN-SHANK and mTOR signaling may result in lasting alteration, for example, in brain connectivity. Indeed, a key outcome of correct connectivity, the establishment of hemisphere lateralization, is altered in PZD mice, which can be observed on a behavioral and molecular level (Grabrucker et al., 2017). At the same time, these results add the establishment of hemisphere dominance and brain lateralization to the list of physiological roles of zinc and zinc-signaling in brain development.

In addition, a reduction in excitatory synapse density but not in the number of inhibitory synapses was observed under zinc-deficient conditions in the striatum and cortex (Grabrucker et al., 2014), which is in line with the hypothesis that an imbalance between excitation and inhibition is a pathomechanism in ASD (Canitano and Pallagrosi, 2017).

Animals with copper overload

Acute copper overload is a crucial feature of Wilson's disease. Therefore, these animals can be considered as animal models for Wilson's disease. Copper overload can be produced by a copper-rich diet or by genetic knockout of critical genes in copper transport such as *Atp7b* (Reed et al., 2018), and occurs in so-called toxic milk mice and the Long-Evans Cinnamon rat, strains bearing a naturally occurring mutation/deletion in *Atb7b*. In contrast to the humans that display motor problems such as tremor, dystonia, gait disturbance, and rigidity and have neuropsychiatric problems such as depression, anxiety, irritability, cognitive impairment, the observed neurological phenotypes are relatively mild in the animal models for Wilson's disease (Reed et al., 2018). However, *Atb7b* mutants show defects in spatial memory and, to a lesser degree, in motor behavior. On a molecular level, this is accompanied by striatal inflammation (Terwel et al., 2011) and high levels of oxidative stress (Wang et al., 2011). Although the expression of genes such as *Snca* (alpha-synuclein, involved in the pathogenesis of Parkinson's disease) and genes related to *App* (amyloid precursor protein, involved in the pathogenesis of Alzheimer's disease) was found altered (Lee et al., 2013), no direct link to ASD was found on the molecular or behavioral level. However, *Commd1*, a gene linked to copper transport that, if dysfunctional, also leads to copper overload, for example, in Bedlington terriers (Vonk et al., 2008), was identified as ASD candidate gene

(Levy et al., 2011). Nevertheless, ASD-like behavior has not been reported in acute copper overload animal models.

Genetic models of copper overload have limitations for studying the effects of excessive maternal copper levels on their offspring. Due to the mutation in *Atp7b* or knockout of the gene, hindering copper export, the milk of these mice is low in copper, which can be lethal for pups already within the first 3 weeks of life unless they are foster-nursed by wildtype mothers (Rauch, 1983; Buiakova et al., 1999). In addition, it is not entirely clear how the mutation affects copper transport across the placenta, and if embryos would develop under excess-copper conditions. The offspring of *Atp7b* knockout mice (maternal copper overload) may have neurological symptoms, including tremor and ataxia (Buiakova et al., 1999), but behavioral assays necessary to detect ASD have rarely been performed.

Apart from the genetic models for copper overload, a diet with high copper concentration can be fed to pregnant mice. As copper transport mechanisms are intact, the embryos developing in pregnant mice with high dietary copper levels may indeed develop under excess-copper conditions. However, excess copper reduces the intestinal absorption of zinc (Bjorklund, 2013). Therefore, pregnant mice with maternal copper overload may absorb less zinc, resulting in a low zinc availability for the developing embryo, and thus, prenatal zinc deficiency. Indeed, the offspring of pregnant mice with dietary copper overload during pregnancy did not show excess copper levels but showed reduced brain zinc levels. In line with the results reported for PZD mice, these pups exposed to prenatal copper overload also had a reduction of synaptic SHANK protein levels and decreased NMDAR subunit 1 concentrations (Baecker et al., 2014). Although their behavior has not been assessed, prenatal copper overload can likely result in prenatal zinc deficiency, reflected in a high Cu/Zn ratio, and thereby contribute to ASD through producing a secondary zinc deficiency that affects the ASD-associated signaling pathways discussed earlier and in Chapter 8. However, an assessment of mice or other animal models with prenatal copper overload regarding ASD-like behavior has not been performed in detail so far.

Animals with an overload of mercury and lead

Acute overload with mercury and lead produces limited behavioral abnormalities in mice. For example, low to moderate doses of methylmercury (MeHg) in young adult mice can impair motor coordination and balance

(Bellum et al., 2007). However, the impact of excess mercury on the behavioral level is much more substantial if the exposure occurs in the pre- and perinatal phase. Although the offspring shows significant alterations in behavior, the mothers of the mice exposed prenatally to MeHg show often no behavioral deficits (Hughes and Annau, 1976).

This is in line with findings in humans, where children exposed to MeHg may exhibit decreased IQ, impaired movements, visuospatial perception, and speech, a medical phenomenon known as fetal Minamata disease (Antunes Dos Santos et al., 2016). In general, in rodent models, the exposure to excessive MeHg leads to, among others, reduced motor activity, a decrease in memory and learning, and depression-like behavior (Bjorklund et al., 2007; Carratu et al., 2006; Dare et al., 2003; Sakamoto et al., 2002; Paletz et al., 2006). In addition, animals exposed to high levels of MeHg prenatally are more susceptible to seizures (Su and Okita, 1976).

Male pups presented more severe impairment of neuromotor performance than female pups (Gandhi et al., 2014) and the male offspring exhibited depression-like behavior (Ceccatelli et al., 2013; Onishchenko et al., 2007) and more severe memory impairments (Onishchenko et al., 2007), revealing a gender effect in mercury overload.

The behavioral abnormalities are reflected by a decreased activity of offspring of animals with mercury overload during pregnancy in the open field assessment (Glover et al., 2009), or decreased exploratory activity (Su and Okita, 1976; Onishchenko et al., 2007). Depressive-like behavior in the MeHg-exposed male offspring was detected in the forced swimming test (Onishchenko et al., 2007). However, some effects of in utero mercury exposure may develop only later in life. Research suggests the existence of a long latency period. In a study using mice, some effects in the open field test, passive avoidance test, and the Morris water maze were observed only 9 months after treatment (Yoshida et al., 2008).

On the molecular level, in utero exposures in postnatal rat and mouse pups show that MeHg induces the altered expression of many genes involved in pathways regulating development such as small GTPase signaling pathways that control cell growth and proliferation (Radonjic et al., 2013) and genes involved in metal/zinc binding and transcription regulation (Glover et al., 2009). The latter results indicate that altered mercury levels will affect zinc homeostasis indirectly. Altered cell proliferation may be the underlying cause for findings of a decreased cell number in the hippocampus and the reduced neural precursor proliferation in pups (Antunes Dos Santos et al., 2016).

In addition, lower expression of brain-derived neurotrophic factor (*Bdnf*) was reported in the hippocampal dentate gyrus along with alterations in DNA methylation, including decreased H3 acetylation at the *Bdnf* promoter IV (Ceccatelli et al., 2013). These results may indicate that in utero exposure to high mercury levels induces long-lasting effects through epigenetic mechanisms (Ceccatelli et al., 2013). Further, the suppressed activity of type III iodothyronine deiodinase, a thyroid hormone-metabolizing enzyme, in the brains of pups from mothers with high levels of mercury was found (Yoshida et al., 2008; Mori et al., 2006). Altered thyroid hormone levels during a critical time window of brain development may also contribute to the long-lasting effects as even transient changes have been reported to result in long-term consequences (Ausó et al., 2004).

Besides, in utero exposure to high mercury levels affects the monoamine and gamma-aminobutyric acid (GABA) neurotransmitter pathways (Yoshida et al., 2008) as well as glutamatergic signaling through and inhibition of glutamate uptake resulting in increased extracellular glutamate levels and responsiveness of NMDA receptors (Antunes Dos Santos et al., 2016).

Although some of the molecular alterations may be linked to ASD, neither high acute nor high pre- and perinatal mercury levels seems to produce an ASD-like behavior. In particular, social impairments seem mostly absent, and repetitive behaviors are not increased (Hill et al., 2015), and communication deficits have not been investigated in detail so far.

Mice with acute lead overload with high concentrations (>500 ppm) starting at weaning (21 days age) and lasting into adulthood show increased anxiety (Soeiro et al., 2007) and impaired olfactory recognition memory (Flores-Montoya et al., 2015), all observed in male but not female mice. Memory impairment was also reported in spatial learning and memory processes using the Morris water maze test (Jett et al., 1997; Toscano and Guilarte, 2005) and a trace fear conditioning paradigm (Anderson et al., 2016). However, similar to mercury, more prominent impairments are reported from experiments that explore neurodevelopmental deficits due to pre- and perinatal lead exposure. Severe and prolonged exposure to high levels of lead results in small brain size of pups with marked retardation of brain maturation (Maker et al., 1975). In general, pre- and perinatal lead overload seems to result in impairments in learning and memory, and attention (Rice, 1990; Chen et al., 1997; Salinas and Huff, 2002; Wang et al., 2016; Anderson et al., 2016). However, abnormal social behavior has also been reported to occur in in utero lead-exposed mice in the three-chamber box test (Hill et al., 2015), and aggressive behavior was significantly increased (Kasten-Jolly et al., 2012). Further, increased anxiety not only results from acute but also developmental exposure to lead (de Souza Lisboa et al., 2005) and may be associated with depression-like behavior. Besides, motor abnormalities have been reported (Leasure et al., 2008). In addition, for pre- and perinatal exposure, lead shows sex-specific effects (Singh et al., 2018).

Coadministration of zinc in animals with high levels of lead and cadmium did not result in significant changes in motor coordination, cognitive activity, and spontaneous motility (Wójcik and Kubajka, 1990), providing a hint that some of the behavioral effects of lead may be caused by affecting the homeostasis of zinc.

On the molecular level, NMDA and AMPA receptor signaling are affected, leading to altered long-term potentiation, whereas GABA signaling seems largely unaffected. Morphological studies showed that lead-exposed animals have a synapse loss (Wang et al., 2016). In addition, gene expression is altered, resulting in, among others, an upregulation of mitogen-activated protein kinases (MAPKs/ERK) signaling, focal adhesion, and vascular endothelial growth factor. Increased expression was also observed for proinflammatory cytokines and chemokine receptors, including IL-7, IL7R, IL-1β, IL-18, and IL-33, while expression of IL-10, an antiinflammatory cytokine, decreased in the cortex (Kasten-Jolly et al., 2012).

Taken together, pre- and perinatal exposure to high levels of lead indeed results in behavioral symptoms associated with ASD, such as impaired social behavior with more aggression and comorbidities associated with ASD, such as cognitive deficits and increased anxiety. The hypothesized underlying mechanisms with a focus on glutamatergic synapses and altered MAPK/ERK signaling, a loss of synapses, and increased inflammatory cytokine expression are the same that are observed in other animal models for ASD and prenatal zinc-deficient mice. However, in utero lead-exposed mice do not display other core features of ASD, such as repetitive behavior, and ultrasonic vocalizations have not been investigated in detail. Intriguingly, some behavioral deficits can be prevented by coadministration of zinc (Wójcik and Kubajka, 1990), which hints at an antagonistic relation of zinc and lead. However, both trace metals show relevance as causative factors for ASD-like behavior in rodent models.

Iron-deficient animals

Rats presenting an acute iron deficiency show neurobehavioral symptoms such as a reversal of the circadian cycles of motor activity, changes in thermoregulation, and stereotyped behavior, as well as an increased pain threshold. In addition, irreversible learning deficits were reported. On the tissue level, these alterations seem to be mostly associated with selective changes in the blood–brain barrier and a decreased function of the dopaminergic system (Yehuda and Youdim, 1989; Yehuda et al., 1986). However, acute iron deficiency was not reported to result in prominent ASD-like behavior.

Pre- and perinatal iron deficiency have lasting effects on brain development and function. Most studies were performed on rats, where iron deficiency is induced during early gestation by feeding pregnant rats an iron-deficient diet until postnatal day 7 or up to postnatal day 21 (Georgieff, 2011). On a gross anatomical level, pre- and perinatal iron deficiency in rodents may result in a decrease in the size of specific brain regions, including the hippocampus (Rao et al., 2011). On the behavioral level, decreased activity, for example, less exploration, and increased anxiety-like behaviors were observed in rats. As described for acute iron-deficient mice, prenatal iron deficiency also targets the dopaminergic system (Beard et al., 2002, 2003). Iron-dependent enzymes, such as tyrosine hydroxylase, the enzyme responsible for catalyzing the conversion of the amino acid tyrosine to L-3,4-dihydroxyphenylalanine, the precursor of dopamine, makes monoamine neurotransmitter metabolism susceptible to iron deficiency (Beard, 2003). However, no prominent ASD-like phenotype has been reported to develop in prenatal iron-deficient animals. In addition, no alterations in major ASD candidate genes or ASD-associated signaling pathways have been reported to be caused by prenatal iron deficiency.

Iron deficiency is known to cause secondary effects on copper metabolism. Iron deficiency results in an increase in maternal liver copper, and copper in maternal blood serum. In addition, the placental copper levels increase (Gambling et al., 2004), which may have effects on the developing embryo. This shows once more that trace metals are in a balance, and alterations in one metal will affect the homeostasis of other trace metals.

Conclusions

Simultaneous alterations in other metal ions likely accompany the alteration of one biometal or the presence of toxic metals. This interconnectedness of trace metals can also be observed in their behavioral phenotypes. For example, the presence of excess lead may result in a displacement of zinc from key zinc-binding proteins and, mouse models for both prenatal zinc

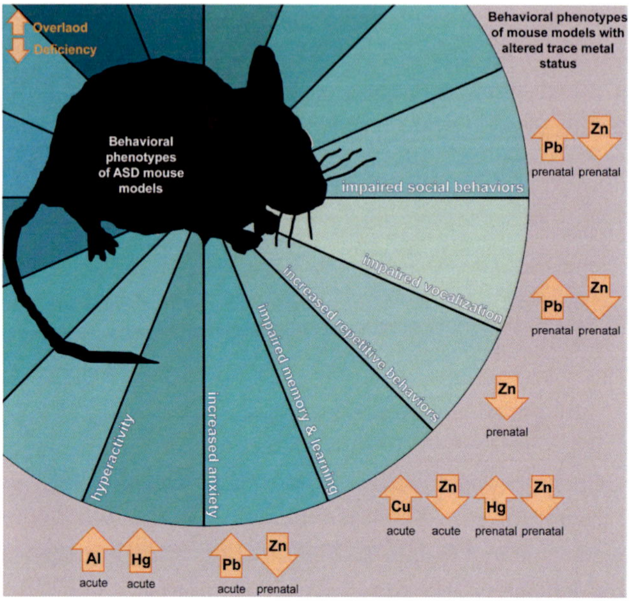

Figure 9.2 Although prenatal zinc-deficient (PZD) mice are best characterized for their ASD-like behavior, some overlap is seen in the behavior of mice with other metal imbalances. These shared behavioral abnormalities could be a result of the imbalance in one metal also affecting other trace metal, including zinc. For example, impaired social behavior and impaired communication that are part of the core features of ASD are not only observed in PZD mice but also mice with prenatal lead overload. However, only PZD mice seem to display abnormal repetitive behaviors. Both acute and zinc-deficient mice have problems with learning and memory, which is also a feature of mice with prenatal mercury overload or acute copper overload. Increased anxiety occurs in PZD mice but is also a result of acute lead exposure. Although PZD mice are not hyperactive, both mice with acute aluminum and mercury overload show increased activity levels. Thus, some links of other trace metals to ASD may be mediated by secondary alterations in zinc.

deficiency and prenatal lead overload display common behavioral impairments (Pfaender and Grabrucker, 2014) (Fig. 9.2).

Thus, for some trace metals that might be codependent, a similar pattern of molecular and behavioral alterations can be found. However, so far, only prenatal zinc deficiency was reported to lead to impairments in all behavioral assays that assess core features and some comorbidities in ASD mouse models, underlining that abnormal zinc biology is a critical factor in the etiology of ASD.

In contrast, the acute zinc deficiency of adult mice does not produce ASD-like behavior. Acute zinc deficiency has been associated with depression-like behavior and changes in inflammatory cytokines as well as altered microbiota composition (Sauer and Grabrucker, 2019). The behavioral differences between acute zinc-deficient mothers and their offspring may be reflected in humans as well. For example, it is known that postpartum depression in women is associated with ASD in their children (Salvanos et al., 2010).

References

Anderson, D.W., Mettil, W., Schneider, J.S., 2016. Effects of low level lead exposure on associative learning and memory in the rat: influences of sex and developmental timing of exposure. Toxicol Lett 246, 57–64.

Antunes Dos Santos, A., Appel Hort, M., Culbreth, M., López-Granero, C., Farina, M., Rocha, J.B., Aschner, M., 2016. Methylmercury and brain development: a review of recent literature. J Trace Elem Med Biol 38, 99–107.

Ausó, E., Lavado-Autric, R., Cuevas, E., Del Rey, F.E., Morreale De Escobar, G., Berbel, P., 2004. A moderate and transient deficiency of maternal thyroid function at the beginning of fetal neocorticogenesis alters neuronal migration. Endocrinology 145 (9), 4037–4047.

Baecker, T., Mangus, K., Pfaender, S., Chhabra, R., Boeckers, T.M., Grabrucker, A.M., 2014. Loss of COMMD1 and copper overload disrupt zinc homeostasis and influence an autism-associated pathway at glutamatergic synapses. Biometals 27 (4), 715–730.

Beard, J., 2003. Iron deficiency alters brain development and functioning. J Nutr 133 (5 Suppl. 1), 1468–1472.

Beard, J.L., Erikson, K.M., Jones, B.C., 2002. Neurobehavioral analysis of developmental iron deficiency in rats. Behav Brain Res 134 (1–2), 517–524.

Beard, J., Erikson, K.M., Jones, B.C., 2003. Neonatal iron deficiency results in irreversible changes in dopamine function in rats. J Nutr 133 (4), 1174–1179.

Bellum, S., Thuett, K.A., Grajeda, R., Abbott, L.C., 2007. Coordination deficits induced in young adult mice treated with methylmercury. Int J Toxicol 26 (2), 115–121.

Bjorklund, G., 2013. The role of zinc and copper in autism spectrum disorders. Acta Neurobiol Exp 73 (2), 225–236.

Bjorklund, O., Kahlstrom, J., Salmi, P., Ogren, S.O., Vahter, M., Chen, J.F., Fredholm, B.B., Dare, E., 2007. The effects of methylmercury on motor activity are sex- and age-dependent, and modulated by genetic deletion of adenosine receptors and caffeine administration. Toxicology 241 (3), 119–133.

Buiakova, O.I., Xu, J., Lutsenko, S., Zeitlin, S., Das, K., Das, S., Ross, B.M., Mekios, C., Scheinberg, I.H., Gilliam, T.C., 1999. Null mutation of the murine ATP7B (Wilson disease) gene results in intracellular copper accumulation and late-onset hepatic nodular transformation. Hum Mol Genet 8 (9), 1665–1671.

Carratu, M.R., Borracci, P., Coluccia, A., Giustino, A., Renna, G., Tomasini, M.C., Raisi, E., Antonelli, T., Cuomo, V., Mazzoni, E., Ferraro, L., 2006. Acute exposure to methylmercury at two developmental windows: focus on neurobehavioral and neurochemical effects in rat offspring. Neuroscience 141 (3), 1619–1629.

Ceccatelli, S., Bose, R., Edoff, K., Onishchenko, N., Spulber, S., 2013. Long-lasting neurotoxic effects of exposure to methylmercury during development. J Intern Med 273 (5), 490–497.

Chen, H.H., Ma, T., Paul, I.A., Spencer, J.L., Ho, I.K., 1997. Developmental lead exposure and two-way active avoidance training alter the distribution of protein kinase C activity in the rat hippocampus. Neurochem Res 22 (9), 1119–1125.

Cole, T.B., Wenzel, H.J., Kafer, K.E., Schwartzkroin, P.A., Palmiter, R.D., 1999. Elimination of zinc from synaptic vesicles in the intact mouse brain by disruption of the ZnT3 gene. Proc Natl Acad Sci USA 96 (4), 1716–1721.
Canitano, R., Pallagrosi, M., 2017. Autism spectrum disorders and schizophrenia spectrum disorders: excitation/inhibition imbalance and developmental trajectories. Front Psychiatr 8, 69.
Crawley, J.N., 2007. Mouse behavioral assays relevant to the symptoms of autism. Brain Pathol 17 (4), 448–459.
Dare, E., Fetissov, S., Hokfelt, T., Hall, H., Ogren, S.O., Ceccatelli, S., 2003. Effects of prenatal exposure to methylmercury on dopamine-mediated locomotor activity and dopamine D2 receptor binding. Naunyn-Schmiedeberg's Arch Pharmacol 367 (5), 500–508.
de Souza Lisboa, S.F., Gonçalves, G., Komatsu, F., Queiroz, C.A., Almeida, A.A., Moreira, E.G., 2005. Developmental lead exposure induces depressive-like behavior in female rats. Drug Chem Toxicol 28 (1), 67–77.
Erickson, J.C., Hollopeter, G., Thomas, S.A., Froelick, G.J., Palmiter, R.D., 1997. Disruption of the metallothionein-III gene in mice: analysis of brain zinc, behavior, and neuron vulnerability to metals, aging, and seizures. J Neurosci 17 (4), 1271–1281.
Flores-Montoya, M.G., Alvarez, J.M., Sobin, C., 2015. Olfactory recognition memory is disrupted in young mice with chronic low-level lead exposure. Toxicol Lett 236 (1), 69–74.
Gambling, L., Dunford, S., McArdle, H.J., 2004. Iron deficiency in the pregnant rat has differential effects on maternal and fetal copper levels. J Nutr Biochem 15 (6), 366–372.
Gandhi, D.N., Panchal, G.M., Dhull, D.K., 2014. Neurobehavioral toxicity in progeny of rat mothers exposed to methylmercury during gestation. Ann Ist Super Sanita 50 (1), 28–37.
Georgieff, M.K., 2011. Long-term brain and behavioral consequences of early iron deficiency. Nutr Rev 69 (Suppl. 1), 43–48.
Glover, C.N., Zheng, D., Jayashankar, S., Sales, G.D., Hogstrand, C., Lundebye, A.K., 2009. Methylmercury speciation influences brain gene expression and behavior in gestationally-exposed mice pups. Toxicol Sci 110 (2), 389–400.
Golub, M.S., Gershwin, M.E., Hurley, L.S., Hendrickx, A.G., Saito, W.Y., 1985. Studies of marginal zinc deprivation in rhesus monkeys: infant behavior. Am J Clin Nutr 42 (6), 1229–1239.
Golub, M.S., Gershwin, M.E., Vijayan, V.K., 1983. Passive avoidance performance of mice fed marginally or severely zinc deficient diets during post-embryonic brain development. Physiol Behav 30 (3), 409–413.
Grabrucker, A.M., 2014. A role for synaptic zinc in ProSAP/Shank PSD scaffold malformation in autism spectrum disorders. Dev Neurobiol 74 (2), 136–146.
Grabrucker, A.M., Knight, M.J., Proepper, C., Bockmann, J., Joubert, M., Rowan, M., Nienhaus, G.U., Garner, C.C., Bowie, J.U., Kreutz, M.R., Gundelfinger, E.D., Boeckers, T.M., 2011. Concerted action of zinc and ProSAP/Shank in synaptogenesis and synapse maturation. EMBO J 30 (3), 569–581.
Grabrucker, S., Boeckers, T.M., Grabrucker, A.M., 2016. Gender dependent evaluation of autism like behavior in mice exposed to prenatal zinc deficiency. Front Behav Neurosci 10, 37.
Grabrucker, S., Haderspeck, J.C., Sauer, A.K., Kittelberger, N., Asoglu, H., Abaei, A., Rasche, V., Schön, M., Boeckers, T.M., Grabrucker, A.M., 2017. Brain lateralization in mice is associated with zinc signaling and altered in prenatal zinc deficient mice that display features of autism spectrum disorder. Front Mol Neurosci 10, 450.
Grabrucker, S., Jannetti, L., Eckert, M., Gaub, S., Chhabra, R., Pfaender, S., Mangus, K., Reddy, P.P., Rankovic, V., Schmeisser, M.J., Kreutz, M.R., Ehret, G., Boeckers, T.M.,

Grabrucker, A.M., 2014. Zinc deficiency dysregulates the synaptic ProSAP/Shank scaffold and might contribute to autism spectrum disorders. Brain 137 (Pt 1), 137–152.
Halas, E.S., Eberhardt, M.J., Diers, M.A., Sandstead, H.H., 1983. Learning and memory impairment in adult rats due to severe zinc deficiency during lactation. Physiol Behav 30 (3), 371–381.
Halas, E.S., Hunt, C.D., Eberhardt, M.J., 1986. Learning and memory disabilities in young adult rats from mildly zinc deficient dams. Physiol Behav 37 (3), 451–458.
Halas, E.S., Sandstead, H.H., 1975. Some effects of prenatal zinc deficiency on behavior of the adult rat. Pediatr Res 9 (2), 94–97.
Hagmeyer, S., Haderspeck, J.C., Grabrucker, A.M., 2014. Behavioral impairments in animal models for zinc deficiency. Front Behav Neurosci 8, 443.
Hill, D.S., Cabrera, R., Wallis Schultz, D., Zhu, H., Lu, W., Finnell, R.H., Wlodarczyk, B.J., 2015. Autism-like behavior and epigenetic changes associated with autism as consequences of in utero exposure to environmental pollutants in a mouse model. Behav Neurol 2015, 426263.
Hughes, J.A., Annau, Z., 1976. Postnatal behavioral effects in mice after prenatal exposure to methylmercury. Pharmacol Biochem Behav 4 (4), 385–391.
Jett, D.A., Kuhlmann, A.C., Farmer, S.J., Guilarte, T.R., 1997. Age-dependent effects of developmental lead exposure on performance in the Morris water maze. Pharmacol Biochem Behav 57 (1–2), 271–279.
Kas, M.J., Glennon, J.C., Buitelaar, J., Ey, E., Biemans, B., Crawley, J., Ring, R.H., Lajonchere, C., Esclassan, F., Talpos, J., Noldus, L.P., Burbach, J.P., Steckler, T., 2014. Assessing behavioural and cognitive domains of autism spectrum disorders in rodents: current status and future perspectives. Psychopharmacology 231 (6), 1125–1146.
Kasten-Jolly, J., Pabello, N., Bolivar, V.J., Lawrence, D.A., 2012. Developmental lead effects on behavior and brain gene expression in male and female BALB/cAnNTac mice. Neurotoxicology 33 (5), 1005–1020.
Koumura, A., Kakefuda, K., Honda, A., Ito, Y., Tsuruma, K., Shimazawa, M., Uchida, Y., Hozumi, I., Satoh, M., Inuzuka, T., Hara, H., 2009. Metallothionein-3 deficient mice exhibit abnormalities of psychological behaviors. Neurosci Lett 467 (1), 11–14.
Leasure, J.L., Giddabasappa, A., Chaney, S., Johnson, J.E., Pothakos, K., Lau, Y.S., Fox, D.A., 2008. Low-level human equivalent gestational lead exposure produces sex-specific motor and coordination abnormalities and late-onset obesity in year-old mice. Environ Health Perspect 116 (3), 355–361.
Lee, B.H., Kim, J.H., Kim, J.M., Heo, S.H., Kang, M., Kim, G.H., Choi, J.H., Yoo, H.W., 2013. The early molecular processes underlying the neurological manifestations of an animal model of Wilson's disease. Metall 5 (5), 532–540.
Levy, D., Ronemus, M., Yamrom, B., Lee, Y.H., Leotta, A., Kendall, J., Marks, S., Lakshmi, B., Pai, D., Ye, K., Buja, A., Krieger, A., Yoon, S., Troge, J., Rodgers, L., Iossifov, I., Wigler, M., 2011. Rare de novo and transmitted copy-number variation in autistic spectrum disorders. Neuron 70 (5), 886–897.
Lokken, P.M., Halas, E.S., Sandstead, H.H., 1973. Influence of zinc deficiency on behavior. Proc Soc Exp Biol Med 144 (2), 680–682.
Maker, H.S., Lehrer, G.M., Silides, D.J., 1975. The effect of lead on mouse brain development. Environ Res 10 (1), 76–91.
Mori, K., Yoshida, K., Hoshikawa, S., Ito, S., Yoshida, M., Satoh, M., Watanabe, C., 2006. Effects of perinatal exposure to low doses of cadmium or methylmercury on thyroid hormone metabolism in metallothionein-deficient mouse neonates. Toxicology 228 (1), 77–84.
Onishchenko, N., Tamm, C., Vahter, M., Hökfelt, T., Johnson, J.A., Johnson, D.A., Ceccatelli, S., 2007. Developmental exposure to methylmercury alters learning and induces depression-like behavior in male mice. Toxicol Sci 97 (2), 428–437.

Paletz, E.M., Craig-Schmidt, M.C., Newland, M.C., 2006. Gestational exposure to methylmercury and n-3 fatty acids: effects on high- and low-rate operant behavior in adulthood. Neurotoxicol Teratol 28 (1), 59–73.
Peters, D.P., 1978. Effects of prenatal nutritional deficiency on affiliation and aggression in rats. Physiol Behav 20 (4), 359–362.
Pfaender, S., Grabrucker, A.M., 2014. Characterization of biometal profiles in neurological disorders. Metallomics 6 (5), 960–977.
Radonjic, M., Cappaert, N.L., de Vries, E.F., de Esch, C.E., Kuper, F.C., van Waarde, A., Dierckx, R.A., Wadman, W.J., Wolterbeek, A.P., Stierum, R.H., de Groot, D.M., 2013. Delay and impairment in brain development and function in rat offspring after maternal exposure to methylmercury. Toxicol Sci 133 (1), 112–124.
Rao, R., Tkac, I., Schmidt, A.T., Georgieff, M.K., 2011. Fetal and neonatal iron deficiency causes volume loss and alters the neurochemical profile of the adult rat hippocampus. Nutr Neurosci 14 (2), 59–65.
Rauch, H., 1983. Toxic milk, a new mutation affecting cooper metabolism in the mouse. J Hered 74 (3), 141–144.
Reed, E., Lutsenko, S., Bandmann, O., 2018. Animal models of Wilson disease. J Neurochem 146 (4), 356–373.
Rice, D.C., 1990. Lead-induced behavioral impairment on a spatial discrimination reversal task in monkeys exposed during different periods of development. Toxicol Appl Pharmacol 106 (2), 327–333.
Sakamoto, M., Kakita, A., Wakabayashi, K., Takahashi, H., Nakano, A., Akagi, H., 2002. Evaluation of changes in methylmercury accumulation in the developing rat brain and its effects: a study with consecutive and moderate dose exposure throughout gestation and lactation periods. Brain Res 949 (1–2), 51–59.
Salinas, J.A., Huff, N.C., 2002. Lead and conditioned fear to contextual and discrete cues. Neurotoxicol Teratol 24 (4), 541–550.
Salvanos, P., Sylven, S.M., Papathoma, E., Petridou, E.T., Skalkidou, A., 2010. P01-255 – maternal postpartum depression in association with autistic traits in the offspring. Eur Psychiatr 25 (1), 466.
Sandstead, H.H., Strobel, D.A., Logan, G.M., Marks, E.O., Jacob, R.A., 1978. Zinc deficiency in pregnant rhesus monkeys: effects on behavior of infants. Am J Clin Nutr 31 (5), 844–849.
Sauer, A.K., Grabrucker, A.M., 2019. Zinc deficiency during pregnancy leads to altered microbiome and elevated inflammatory markers in mice. Front Neurosci 13, 1295.
Schoen, M., Asoglu, H., Bauer, H.F., Müller, H.P., Abaei, A., Sauer, A.K., Zhang, R., Song, T.J., Bockmann, J., Kassubek, J., Rasche, V., Grabrucker, A.M., Boeckers, T.M., 2019. SHANK3Transgenic and prenatal zinc-deficient autism mouse models show convergent and individual alterations of brain structures in MRI. Front Neural Circ 13, 6.
Silverman, JL, Yang, M, Lord, C, Crawley, JN, 2010. Behavioural phenotyping assays for mouse models of autism. Nat Rev Neurosci 11 (7), 490–502.
Singh, G., Singh, V., Sobolewski, M., Cory-Slechta, D.A., Schneider, J.S., 2018. Sex-dependent effects of developmental lead exposure on the brain. Front Genet 9, 89.
Soeiro, A.C., Gouvêa, T.S., Moreira, E.G., 2007. Behavioral effects induced by subchronic exposure to Pb and their reversion are concentration and gender dependent. Hum Exp Toxicol 26 (9), 733–739.
Su, M.Q., Okita, G.T., 1976. Behavioral effects on the progeny of mice treated with methylmercury. Toxicol Appl Pharmacol 38 (1), 195–205.
Tahmasebi Boroujeni, S., Naghdi, N., Shahbazi, M., Farrokhi, A., Bagherzadeh, F., Kazemnejad, A., Javadian, M., 2009. The effect of severe zinc deficiency and zinc supplement on spatial learning and memory. Biol Trace Elem Res 130 (1), 48–61.

Tao-Cheng, J.H., Toy, D., Winters, C.A., Reese, T.S., Dosemeci, A., 2016. Zinc stabilizes Shank3 at the postsynaptic density of hippocampal synapses. PLoS One 11 (5), e0153979.

Tassabehji, N.M., Corniola, R.S., Alshingiti, A., Levenson, C.W., 2008. Zinc deficiency induces depression-like symptoms in adult rats. Physiol Behav 95 (3), 365–369.

Terwel, D., Löschmann, Y.N., Schmidt, H.H., Schöler, H.R., Cantz, T., Heneka, M.T., 2011. Neuroinflammatory and behavioural changes in the Atp7B mutant mouse model of Wilson's disease. J Neurochem 118 (1), 105–112.

Toscano, C.D., Guilarte, T.R., 2005. Lead neurotoxicity: from exposure to molecular effects. Brain Res Brain Res Rev 49 (3), 529–554.

Vonk, W.I., Wijmenga, C., van de Sluis, B., 2008. Relevance of animal models for understanding mammalian copper homeostasis. Am J Clin Nutr 88 (3), 840–845.

Wang, J., Yuan, B., Guerrero, C., Bahde, R., Gupta, S., Wang, Y., 2011. Quantification of oxidative DNA lesions in tissues of Long-Evans Cinnamon rats by capillary high-performance liquid chromatography-tandem mass spectrometry coupled with stable isotope-dilution method. Anal Chem 83 (6), 2201–2209.

Wang, T., Guan, R.L., Liu, M.C., Shen, X.F., Chen, J.Y., Zhao, M.G., Luo, W.J., 2016. Lead exposure impairs Hippocampus related learning and memory by altering synaptic plasticity and morphology during juvenile period. Mol Neurobiol 53 (6), 3740–3752.

Wójcik, A., Kubajka, J., 1990. Experimental study of heavy metal interactions in mice. Ann Univ Mariae Curie Sklodowska Med 45, 21–26.

Yehuda, S., Youdim, M.E., Mostofsky, D.I., 1986. Brain iron-deficiency causes reduced learning capacity in rats. Pharmacol Biochem Behav 25 (1), 141–144.

Yehuda, S., Youdim, M.B., 1989. Brain iron: a lesson from animal models. Am J Clin Nutr 50 (3 Suppl. l), 618–625 discussion 625-9.

Yoo, M.H., Kim, T.Y., Yoon, Y.H., Koh, J.Y., 2016. Autism phenotypes in ZnT3 null mice: involvement of zinc dyshomeostasis, MMP-9 activation and BDNF upregulation. Sci Rep 6, 28548.

Yoshida, M., Shimizu, N., Suzuki, M., Watanabe, C., Satoh, M., Mori, K., Yasutake, A., 2008. Emergence of delayed methylmercury toxicity after perinatal exposure in metallothionein-null and wild-type C57BL mice. Environ Health Perspect 116 (6), 746–751.

Yu, X., Jin, L., Zhang, X., Yu, X., 2013. Effects of maternal mild zinc deficiency and zinc supplementation in offspring on spatial memory and hippocampal neuronal ultrastructural changes. Nutrition 29 (2), 457–461.

CHAPTER 10

Animal models for autism—links to biometal abnormalities

Contents

Several studies have shown that the pathogenesis of ASD has a strong genetic component (Delorme et al., 2013), but each risk-associated genetic variant is rare in the ASD population, meaning that no single genetic factor has been able to account for even 10% of all cases of ASD. However, investigation of the identified candidate genes showed a clear enrichment in genes coding for proteins that are found at synapses in the central nervous system (CNS) (Kelleher et al., 2012). Among these are adhesion proteins of the neuroligin and neurexin families (Jamain et al., 2003), SHANK (SH3 and multiple ankyrin repeat domains) scaffold proteins of the postsynaptic density (PSD) (Grabrucker et al., 2011), and proteins of the mTOR (mechanistic target of rapamycin) signaling pathway (Huber et al., 2015). Together, these and further ASD-associated proteins form a signaling network. The disturbance of this network may be the underlying cause of a large share of ASD cases although different proteins in the network may be affected in different individuals. Interestingly, critical proteins of this signaling network are zinc binding and regulated by zinc signaling and local zinc concentrations (Grabrucker, 2014).

However, if an impairment of this synaptic protein network is the molecular biological correlate of ASD, the pathomechanisms of other risk factors, genetic or nongenetic, may converge on inducing alterations in this network. As a consequence, zinc signaling as the principal regulator of the network should modify the phenotype of several animal models for ASD, despite ASD being caused by different factors in the animals. Indeed, zinc has been reported to modify the phenotype of rodent models for ASD

Biometals in Autism Spectrum Disorders
ISBN 978-0-12-821132-8
https://doi.org/10.1016/B978-0-12-821132-8.00010-0

caused by deletion (knockout (KO)) of *Shank2*, *Shank3*, maternal immune activation (MIA), and in utero exposure to valproic acid (VPA).

Shank2 and *Shank3* KO mice

Shank2 and *Shank3* KO mice have been subject to several studies that all confirmed the presence of ASD-like behavior in these animals and aimed to identify the molecular consequences of the loss of SHANK2 and SHANK3 concerning ASD (Won et al., 2012; Schmeisser et al., 2012; Bozdagi et al., 2010; Yang et al., 2012; Wang et al., 2011; Peça et al., 2011). A key feature of both proteins is that they are recruited to the PSD of excitatory synapses through their C-terminal zinc-binding SAM domains (Grabrucker et al., 2011).

Human individuals with ASD-associated mutations in *SHANK2* and *SHANK3*, or a loss of *SHANK3* that may result in Phelan McDermid syndrome, are heterozygous, meaning that they still possess one intact copy of the gene. Given that SHANK2 and SHANK3 occur in a soluble and PSD-bound state and that additional zinc will shift the balance toward the PSD-bound state, zinc supplementation may recruit remaining soluble protein or more of nonmutated or deleted SHANK family members to the PSD, and thereby lead to an amelioration of the ASD phenotype, a model proposed by Hagmeyer et al. (2018) (Fig. 10.1). Indeed, this model was later confirmed experimentally. Increasing levels of dietary zinc were able to alter behavioral and synaptic deficits in *Shank3* knockout mice (Shank3ex13-1616−/−). Six weeks of zinc supplementation prevented ASD-related repetitive and anxiety behaviors and deficits in social novelty recognition. On the molecular level, dietary zinc supplementation indeed increased SHANK2 levels at synapses and altered long-term potentiation at cortico-striatal synapses (Fourie et al., 2018).

Apart from dietary zinc supplementation, local zinc signaling in the brain can be influenced more directly by zinc ionophores. Among these, clioquinol (CQ) (iodochlorhydroxyquin) has shown promising effects in preclinical studies exploring CQ as a treatment of Alzheimer's disease (Bush, 2002). CQ is a moderate zinc chelator. A possible mode of action is that CQ, due to is cell permeability, can bind free zinc, for example, from presynaptic zinc-containing neurotransmitter vesicles. However, CQ has a lower affinity for zinc than many proteins. After diffusing from the vesicle to the cytoplasm, CQ may act as a donor of zinc for proteins with higher affinity. Thus, instead of removing zinc from the cell, CQ redistributes zinc

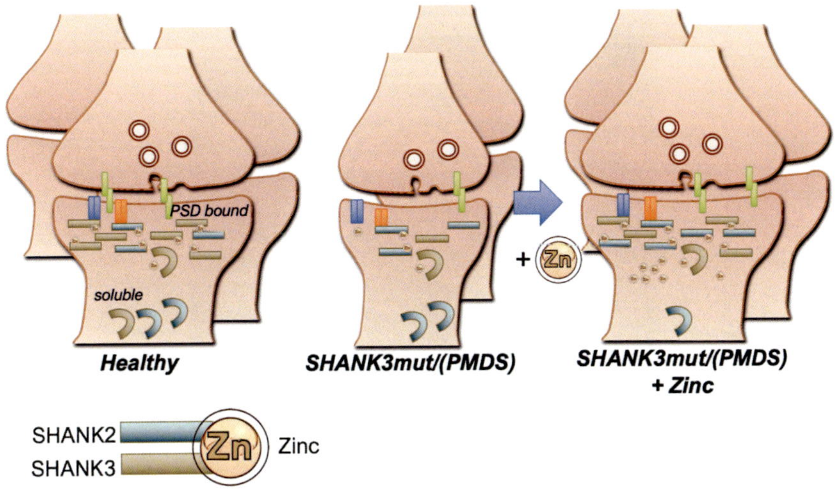

Figure 10.1 Proposed model for the mode of action of zinc supplementation in *Shank3*-deficient conditions. Left: A soluble pool of SHANK2 and SHANK3 proteins exists at the postsynapse, as well as a PSD-bound pool. Recruitment of SHANKs into the PSD occurs through zinc-signaling. By strengthening the synaptic scaffold, active (zinc releasing) synapses will mature and be maintained. Via *trans*-synaptic neuroligin and neurexin complexes, the level of PSD-bound SHANK3 coordinates structural changes of the PSD with the presynapse. Middle: Loss of SHANK3 due to heterozygous deletion or mutation destabilizes the SHANK scaffold, which results in impaired formation and maturation of synapses visible through a lower synapse density in SHANK2 and SHANK3 knockout mice. Right: Zinc supplementation might restore PSD-bound SHANK levels by recruitment of remaining soluble SHANK3 to the PSD, and additional recruitment of SHANK2 proteins. These effects may strengthen the PSD scaffold enough to allow the establishment of normal synaptic connectivity.

into the cell. This zinc mobilization was shown to improve social interactions in *Shank2* KO mice (Lee et al., 2015) (Fig. 10.2).

MIA rats/mice

MIA has been associated with ASD in epidemiological studies. It can be modeled in mice through injection of a viral mimetic (polyinosinic:polycytidylic acid (poly(I:C)), or lipopolysaccharides (LPS) found in the outer membrane of gram-negative bacteria, into pregnant mice (Hill-Yardin et al., 2017). The resulting maternal immune response was shown to lead to lasting changes in the offspring, resulting, among others, in ASD-like behavioral dysfunction, including social and communication deficits (Bilbo et al., 2018). On the molecular level, altered expression of immune

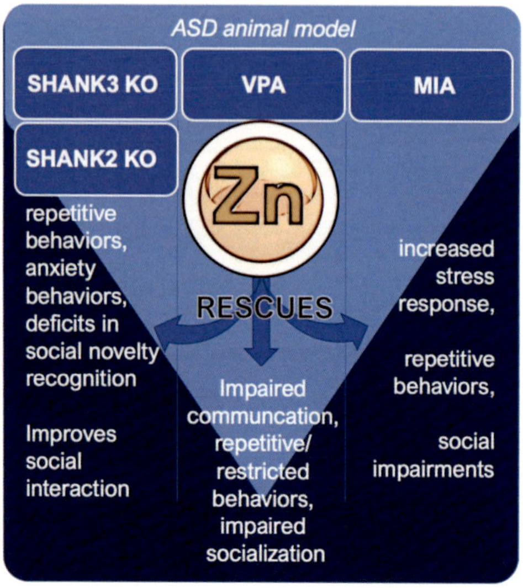

Figure 10.2 Zinc supplementation rescues repetitive behaviors, anxiety, communication deficits, and impaired social behavior (all core features of ASD) in several rodent models for ASD. This hints at a common pathomechanism for the different models that may be centered around zinc deficiency or dysregulation of zinc-dependent proteins (Cezar et al., 2018; Fourie et al., 2018; Kirsten et al., 2015a,b; Lee et al., 2015).

molecules and/or neuroinflammation were suspected to contribute mechanistically to the phenotype of these mice (Tsukada et al., 2019). Through this, other risk factors for ASD, such as exposure to certain toxins, maternal stress, and maternal obesity, may be integrated as causative factors for ASD into the current understanding of the ASD pathomechanisms, as they all impact inflammatory or immune pathways (Bilbo et al., 2018). Increased levels of proinflammatory cytokines and neuroinflammation may again ultimately affect known ASD-associated signaling pathways at synapses in the CNS.

Prenatal treatment of rats with LPS not only induces short- and long-term behavioral and neuroimmune changes in the offspring but also induces hypozincemia (Galvão et al., 2015). Probably, therefore, the behavior of MIA mice and rats resembles that of prenatal zinc-deficient mice. If the behavioral abnormalities of MIA mice are indeed a consequence of prenatal zinc deficiency, prenatal zinc supplementation should prevent those impairments that are caused by low developmental zinc status. In line with this model, prenatal zinc supplementation of mice with MIA reduced the increased stress response in adult rats, hinting at normalization of increased anxiety (Galvão et al., 2015). In addition, prenatal LPS exposure impaired play behaviors (social impairment) and T-maze spontaneous alternations (repetitive behavior/cognitive impairments) (Kirsten et al., 2015a), and reproductive communication (impaired communication) (Kirsten et al., 2015b), as part of the ASD-like phenotype of mice after MIA. Prenatal zinc treatment prevented these ASD-related behavioral impairments (Kirsten et al., 2015a,b) (Fig. 10.2). Thus, rodents with MIA have low zinc status and ASD-like behavior. The core features of ASD that are also seen in prenatal zinc-deficient mice are rescued in animals with MIA by zinc supplementation/prevention of hypozincemia. These results show that altered zinc homeostasis is a key feature of the pathomechanism of MIA.

In line with the current model, that a specific synaptic signaling pathway centered around SHANK proteins and mTOR signaling is critical for the development of ASD, prenatal LPS treatment increased mTOR levels in the striatum. Prenatal zinc treatment prevented these mTOR disturbances in the offspring (Kirsten et al., 2015a). Besides, increased levels of BDNF were detected after prenatal LSP exposure and were reduced by prenatal zinc supplementation (Kirsten et al., 2015b).

VPA rats/mice

VPA is used as an antiepileptic treatment. However, exposure to VPA in utero produces ASD-like behavior in rodents that is visible through impaired ultrasonic vocalization, repetitive behaviors, and impaired socialization (Nicolini and Fahnestock, 2018). The exposure to VPA may alter zinc metabolism resulting in transient zinc deficiency (Cezar et al., 2018). Therefore, as seen in MIA mice, the ASD-like behavior may be caused by altered zinc status of the pregnant mice, and the prevention of hypozincemia should rescue ASD-like behavior.

Prenatal treatment of VPA exposed rats with zinc reduced VPA-induced ASD-like behaviors (Cezar et al., 2018) (Fig. 10.2). As a potential molecular mechanism of VPA-induced ASD-like behavior, a reduction of tyrosine hydroxylase (TH) that has been measured in VPA exposed animals is discussed. TH is an important enzyme in the metabolism of the neurotransmitter dopamine. However, despite the rescue of ASD-like behavior by zinc, no effects were observed on TH activity and expression levels (Cezar et al., 2018). Thus, other signaling pathways may be affected by VPA, including the zinc-dependent SHANK/mTOR signaling processes. Recently, it was shown that rapamycin, a modulator of mTOR signaling, also rescues some ASD-like behaviors of VPA exposed mice (Kotajima-Murakami et al., 2019). Thus, also in VPA exposed rodent models for ASD, prenatal zinc deficiency may be a key contributing causative factor to the observed ASD-like phenotype.

Conclusions

Glutamate signaling and zinc signaling are closely connected. This is especially visible in investigating *Shank2* and *Shank3* knockout mouse models and prenatal zinc-deficient (PZD) mice. Intact SHANK signaling at glutamatergic synapses and the associated regulation of critical signaling pathways seem to be central for healthy synapse development and function, and disruption during a critical time window in brain development that, in mice, occurs already in utero, is causative for the ASD pathology. Therefore, mutations and loss of SHANK2 or SHANK3 are described as major risk factors in the etiology of ASD.

On the other hand, these proteins are regulated by zinc signaling. Intact SHANK2 and SHANK3 proteins are rendered dysfunctional in the absence of sufficient zinc. Consequently, prenatal zinc deficiency results in ASD-like behavior. The dependency of SHANK2 and SHANK3 on zinc is apparent through (1) the loss of SHANK proteins in PZD mice and (2) the ability of zinc supplementation and zinc ionophores to prevent and rescue ASD-like behavior in *Shank2* and *Shank3* knockout mice.

Based on these data, one can speculate that individuals with mutations in genes encoding molecules important for excitatory synaptic signaling (i.e., *SHANK3, FMRP1, TSC1, TSC2, NLGN, NRXN, NF1,* etc.) have a heightened sensitivity to the effects of developmental zinc deficiency. This may also be the reason why individuals with ASD frequently present zinc deficiency not only during pregnancy but also later in life, including individuals with ASD that harbor mutations in essential ASD genes. If zinc availability modulates the ASD phenotype and its severity, as has been shown in mouse models, then high zinc levels may result in an individual with abnormal glutamatergic synaptic signaling to have only mild symptoms that may not lead to an ASD diagnosis. In return, low zinc availability may increase the severity, and therefore, individuals with zinc deficiency are overrepresented, even in those groups of individuals with ASD, where a mutation in synaptic genes is found as a cause for ASD.

Along with zinc deficiency, due to the antagonistic relationship with copper as well as toxic metals like mercury and lead, the Cu/Zn ration will be abnormal in these individuals, and higher levels of toxic metals detected, which seems to be confirmed through epidemiological studies. However, future research needs to investigate whether other ASD mouse models such as neurexin or neuroligin mutant mice also respond to differences in zinc availability. Unfortunately, many animal experiments in the field of ASD research do not consider the influence of zinc status of their mice/rats on the pathology and behavioral phenotype of their test animals. For example, almost no publication even indicated the type of diet the animals received and their zinc content. However, it is common to use so-called fortified diets in a laboratory setting with the idea to provide the animal with all necessary nutrients (vitamins, minerals) to ensure its health. Fortified diets often contain a surplus of vitamins (because vitamins are unstable and some losses occur during autoclaving of the food pellets), but also minerals. Thus, most experimental data generated in this field may be based on animals that are de facto zinc supplemented. Interestingly, while humans with mutations in ASD-associated genes are often heterozygous for the mutation, in experiments with knockout mice, only homozygous knockout mice show a significant ASD-like phenotype. Future research on ASD animal models may have to critically evaluate whether the biometal composition of their diet may influence the results obtained.

If the zinc availability is such a critical factor for the function of the ASD-associated signaling pathway at glutamatergic synapses, then any other genetic or nongenetic factor that influences zinc status during pregnancy should also be a risk factor for ASD. The two most prominent ASD mouse models created through nongenetic manipulation are the offspring of mice with MIA and VPA exposure.

It has been shown that MIA induces alterations in metal homeostasis. LPS treatment reduced maternal levels of zinc, magnesium, and manganese (Kirsten et al., 2015a). Thus, it can be speculated that any factor resulting in MIA will also affect zinc homeostasis. Although the increased expression of proinflammatory cytokines has been discussed as a causative mechanism for ASD after MIA, cytokine release may be a consequence of zinc deficiency, and PZD is again a key mediator of the development of ASD-like behavior. In line with this model, zinc supplementation rescued ASD-like behavior in the offspring of mice with MIA.

Similarly, several mechanisms were proposed, how in utero VPA exposure may result in ASD-like behavior. Intriguingly, VPA binds to zinc (Hurd et al., 1983), and because of the symptoms of children born from mothers with VPA exposure, zinc deficiency has long been suspected of contributing to their phenotype (Zaki et al., 2010). In rodent models, zinc supplementation ameliorated all core symptoms of ASD.

The question therefore arises, whether abnormal zinc metabolism is the central motif that can unify different genetic and nongenetic risk factors for ASD into one pathomechanism (see Chapter 14).

References

Bilbo, S.D., Block, C.L., Bolton, J.L., Hanamsagar, R., Tran, P.K., 2018. Beyond infection – maternal immune activation by environmental factors, microglial development, and relevance for autism spectrum disorders. Exp Neurol 299 (Pt A), 241–251.

Bozdagi, O., Sakurai, T., Papapetrou, D., Wang, X., Dickstein, D.L., Takahashi, N., Kajiwara, Y., Yang, M., Katz, A.M., Scattoni, M.L., Harris, M.J., Saxena, R., Silverman, J.L., Crawley, J.N., Zhou, Q., Hof, P.R., Buxbaum, J.D., 2010. Haploinsufficiency of the autism-associated Shank3 gene leads to deficits in synaptic function, social interaction, and social communication. Mol Autism 1 (1), 15.

Bush, A.I., 2002. Metal complexing agents as therapies for Alzheimer's disease. Neurobiol Aging 23 (6), 1031–1038.

Cezar, L.C., Kirsten, T.B., da Fonseca, C.C.N., de Lima, A.P.N., Bernardi, M.M., Felicio, L.F., 2018. Zinc as a therapy in a rat model of autism prenatally induced by valproic acid. Prog Neuropsychopharmacol Biol Psychiatry 84 (Pt A), 173–180.

Delorme, R., Ey, E., Toro, R., Leboyer, M., Gillberg, C., Bourgeron, T., 2013. Progress toward treatments for synaptic defects in autism. Nat Med 19 (6), 685–694.

Fourie, C., Vyas, Y., Lee, K., Jung, Y., Garner, C.C., Montgomery, J.M., 2018. Dietary zinc supplementation prevents autism related behaviors and striatal synaptic dysfunction in Shank3 Exon 13-16 mutant mice. Front Cell Neurosci 12, 374.

Galvão, M.C., Chaves-Kirsten, G.P., Queiroz-Hazarbassanov, N., Carvalho, V.M., Bernardi, M.M., Kirsten, T.B., 2015. Prenatal zinc reduces stress response in adult rat offspring exposed to lipopolysaccharide during gestation. Life Sci 120, 54–60.

Grabrucker, A.M., 2014. A role for synaptic zinc in ProSAP/Shank PSD scaffold malformation in autism spectrum disorders. Dev Neurobiol 74 (2), 136–146.

Grabrucker, A.M., Schmeisser, M.J., Schoen, M., Boeckers, T.M., 2011. Postsynaptic ProSAP/Shank scaffolds in the cross-hair of synaptopathies. Trends Cell Biol 21 (10), 594–603.

Hagmeyer, S., Sauer, A.K., Grabrucker, A.M., 2018. Prospects of zinc supplementation in autism spectrum disorders and Shankopathies such as Phelan McDermid syndrome. Front Synaptic Neurosci 10, 11.

Hill-Yardin, E.L., McKeown, S.J., Novarino, G., Grabrucker, A.M., 2017. Extracerebral dysfunction in animal models of autism spectrum disorder. Adv Anat Embryol Cell Biol 224, 159–187.

Huber, K.M., Klann, E., Costa-Mattioli, M., Zukin, R.S., 2015. Dysregulation of mammalian target of rapamycin signaling in mouse models of autism. J Neurosci 35 (41), 13836–13842.

Hurd, R.W., Wilder, B.J., Van Rinsvelt, H.A., 1983. Valproate, birth defects, and zinc. Lancet 1 (8317), 181.

Jamain, S., Quach, H., Betancur, C., Råstam, M., Colineaux, C., Gillberg, I.C., Soderstrom, H., Giros, B., Leboyer, M., Gillberg, C., Bourgeron, T., 2003. Mutations of the X-linked genes encoding neuroligins NLGN3 and NLGN4 are associated with autism. Nat Genet 34 (1), 27–29.

Kelleher, R.J., Geigenmüller, U., Hovhannisyan, H., Trautman, E., Pinard, R., Rathmell, B., Carpenter, R., Margulies, D., 2012. High-throughput sequencing of mGluR signaling pathway genes reveals enrichment of rare variants in autism. PLoS One 7 (4), e35003.
Kirsten, T.B., Chaves-Kirsten, G.P., Bernardes, S., Scavone, C., Sarkis, J.E., Bernardi, M.M., Felicio, L.F., 2015a. Lipopolysaccharide exposure induces maternal hypozincemia, and prenatal zinc treatment prevents autistic-like behaviors and disturbances in the striatal dopaminergic and mTOR systems of offspring. PLoS One 10 (7), e0134565.
Kirsten, T.B., Queiroz-Hazarbassanov, N., Bernardi, M.M., Felicio, L.F., 2015b. Prenatal zinc prevents communication impairments and BDNF disturbance in a rat model of autism induced by prenatal lipopolysaccharide exposure. Life Sci 130, 12–17.
Kotajima-Murakami, H., Kobayashi, T., Kashii, H., Sato, A., Hagino, Y., Tanaka, M., Nishito, Y., Takamatsu, Y., Uchino, S., Ikeda, K., 2019. Effects of rapamycin on social interaction deficits and gene expression in mice exposed to valproic acid in utero. Mol Brain 12 (1), 3.
Lee, E.J., Lee, H., Huang, T.N., Chung, C., Shin, W., Kim, K., Koh, J.Y., Hsueh, Y.P., Kim, E., 2015. Trans-synaptic zinc mobilization improves social interaction in two mouse models of autism through NMDAR activation. Nat Commun 6, 7168.
Nicolini, C., Fahnestock, M., 2018. The valproic acid-induced rodent model of autism. Exp Neurol 299 (Pt A), 217–227.
Peça, J., Feliciano, C., Ting, J.T., Wang, W., Wells, M.F., Venkatraman, T.N., Lascola, C.D., Fu, Z., Feng, G., 2011. Shank3 mutant mice display autistic-like behaviours and striatal dysfunction. Nature 472 (7344), 437–442.
Schmeisser, M.J., Ey, E., Wegener, S., Bockmann, J., Stempel, A.V., Kuebler, A., Janssen, A.L., Udvardi, P.T., Shiban, E., Spilker, C., Balschun, D., Skryabin, B.V., Dieck, S., Smalla, K.H., Montag, D., Leblond, C.S., Faure, P., Torquet, N., Le Sourd, A.M., Toro, R., Grabrucker, A.M., Shoichet, S.A., Schmitz, D., Kreutz, M.R., Bourgeron, T., Gundelfinger, E.D., Boeckers, T.M., 2012. Autistic-like behaviours and hyperactivity in mice lacking ProSAP1/Shank2. Nature 486 (7402), 256–260.
Tsukada, T., Shimada, H., Sakata-Haga, H., Iizuka, H., Hatta, T., 2019. Molecular mechanisms underlying the models of neurodevelopmental disorders in maternal immune activation relevant to the placenta. Congenit Anom (Kyoto) 59 (3), 81–87.
Wang, X., McCoy, P.A., Rodriguiz, R.M., Pan, Y., Je, H.S., Roberts, A.C., Kim, C.J., Berrios, J., Colvin, J.S., Bousquet-Moore, D., Lorenzo, I., Wu, G., Weinberg, R.J., Ehlers, M.D., Philpot, B.D., Beaudet, A.L., Wetsel, W.C., Jiang, Y.H., 2011. Synaptic dysfunction and abnormal behaviors in mice lacking major isoforms of Shank3. Hum Mol Genet 20 (15), 3093–3108.
Won, H., Lee, H.R., Gee, H.Y., Mah, W., Kim, J.I., Lee, J., Ha, S., Chung, C., Jung, E.S., Cho, Y.S., Park, S.G., Lee, J.S., Lee, K., Kim, D., Bae, Y.C., Kaang, B.K., Lee, M.G., Kim, E., 2012. Autistic-like social behaviour in Shank2-mutant mice improved by restoring NMDA receptor function. Nature 486 (7402), 261–265.
Yang, M., Bozdagi, O., Scattoni, M.L., Wöhr, M., Roullet, F.I., Katz, A.M., Abrams, D.N., Kalikhman, D., Simon, H., Woldeyohannes, L., Zhang, J.Y., Harris, M.J., Saxena, R., Silverman, J.L., Buxbaum, J.D., Crawley, J.N., 2012. Reduced excitatory neurotransmission and mild autism-relevant phenotypes in adolescent Shank3 null mutant mice. J Neurosci 32 (19), 6525–6541.
Zaki, S.A., Phulsundar, A., Shanbag, P., Mauskar, A., 2010. Fetal valproate syndrome in a 2-month-old male infant. Ann Saudi Med 30 (3), 233–235.

CHAPTER 11

Human stem cell models linking biometal abnormalities and autism

Contents

iPSCs (induced pluripotent stem cells) are cells that are derived from adult somatic cells and have regained the capacity to differentiate into any type of cell found in the body (pluripotency) through genetic reprogramming. This reprogramming occurs through artificial expression of the four key genes Oct3/4, Sox2, Klf4, and c-Myc, the so-called Yamanaka factors, named after Shinya Yamanaka who first described the processes (Takahashi and Yamanaka, 2006). iPSCs can differentiate into all three germ layers. Following the initial discovery of the technology that produces iPSCs, protocols have been developed that allow differentiating iPSCs into organ-specific tissue types. Importantly, cells can be sourced from individuals with ASD and reprogrammed to iPSCs, and subsequently differentiated into desired cell types such as neurons (Park et al., 2008). The cell-type-specific differentiation allows investigating not only the mature cells resulting from iPSC differentiation but also intermediate stages that, at least in part, recapitulate developmental processes. Although iPSCs have been frequently employed so far to understand ASD-associated mutations and their pathomechanisms in different tissues (Russo et al., 2019), they have rarely been used to investigate the influence of trace metals on differentiation and resulting cells, nor has trace metal biology been investigated in detail in cells harboring mutations in ASD candidate genes. However, few studies have been conducted that revealed new insights into the interplay between ASD and trace metals, in particular, zinc homeostasis.

Biometals in Autism Spectrum Disorders
ISBN 978-0-12-821132-8
https://doi.org/10.1016/B978-0-12-821132-8.00011-2

Human stem cells to model neuronal trace metal biology in health and ASD

Human embryonic stem cells (hESC) were used to investigate the effects of methyl-mercury and valproic acid (VPA) on developing neurons (Krug et al., 2013). Although VPA triggered a high number of transcriptional changes, the effects of methyl-mercury were less dramatic. Interestingly, a relatively large "common response" to both treatments was observed. Among the genes regulated by both VPA and methyl-mercury exposure were, for example, the ASD-associated genes *HIF-1* (*SETD2*, SET Domain Containing 2) (O'Roak et al., 2012) and *FOXP1* (Forkhead Box P1) (O'Roak et al., 2011).

iPSCs from healthy individuals were investigated to understand the expression of zinc transporters during neuronal differentiation (Pfaender et al., 2016). Several zinc transporters show a developmental and differentiation stage-dependent expression or altered expression levels during the differentiation process from iPSC to mature motor neurons. For example, *ZnT2* (*SLC30A2*) was only expressed in neural stem cells, but not in iPSC, and also not in mature neurons. In addition, a steady increase in the expression levels of *ZnT4* (*SLC30A4*) from iPSC mature motor neurons was seen. Similarly, *ZnT1* (*SLC30A1*), *ZnT3* (*SLC30A3*), and *ZnT6* (*SLC30A6*) levels were significantly higher in motor neurons compared to iPSC.

In contrast, *ZnT5* (*SLC30A5*), *ZIP1* (*SLC39A1*), and *ZIP3* (*SLC39A3*) mRNA were expressed at a similar level at all stages of motor neuron differentiation. These results indicate that the zinc levels in the cytoplasm and intracellular vesicles/organelles are tightly controlled during different stages of neuronal differentiation. Thus, zinc signaling may play a role in orchestrating the differentiation process. In line with this hypothesis, repeating the differentiation process under low-zinc conditions shows that cellular mechanisms try to compensate for the low availability of zinc. For example, *ZnT1* (*SLC30A1*) expression (the major zinc exporter of cells) was significantly reduced in mature zinc-depleted motor neurons to minimize cellular losses of zinc. In contrast, the expression of *ZnT5* (*SLC30A5*) that may transport zinc from stores in the Golgi apparatus to the cytoplasm was upregulated, along with *ZIP1* (*SLC39A1*), a zinc importer and *MT3*, intracellular zinc-binding proteins (Pfaender et al., 2016). However, in zinc-depleted conditions, where compensatory regulation of zinc homeostasis genes is ineffective in protecting against zinc deficiency, zinc deficiency triggers proapoptotic processes very early during neuronal differentiation. More importantly, the morphology and functionality of the resulting mature iPSC-derived motor neurons are affected by altered zinc signaling during neuronal differentiation. For example, the cell size is reduced, and glutamatergic and GABAergic synaptic transmission impaired, in contrast to acetylcholinergic signaling that remained unaffected (Pfaender et al., 2016). Thus, an adequate intracellular zinc status during human neuronal development ensures neuronal survival and is necessary for functional glutamatergic and GABAergic neurotransmission.

iPSC to model gastrointestinal zinc absorption in ASD

Gastrointestinal (GI) problems are frequent comorbidity of ASD (Horvath and Perman, 2002). The GI problems result in, among others, inflammation of the upper and lower intestinal tract, increased intestinal permeability ("leaky gut"), and decreased digestive enzyme activities. To understand the etiology of the GI abnormalities in ASD, and whether there is a link to the reported alterations in trace metal status of individuals with ASD, and in particular to zinc deficiency, iPS cells from patients with mutations/deletions in *SHANK3* were differentiated to functional epithelial cells (enterocytes) of the GI tract (Pfaender et al., 2017).

SHANK3 is a major ASD-associated gene, and mutations and deletions have been reported in many individuals with ASD (Leblond et al., 2014). In addition, loss of, or deletions within the *SHANK3* gene are likely contributing to ASD in Phelan McDermid syndrome (PMDS). Individuals with PMDS also frequently have GI problems (Kolevzon et al., 2014). In addition, a high incidence of zinc deficiency has been reported in PMDS (Grabrucker et al., 2014; Pfaender et al., 2017).

The analysis of iPSC-derived enterocytes from PMDS patients revealed that they express *SHANK3* on mRNA and protein level. Subsequently, in expression analysis of gut epithelium cells of mice, the expression of several more ASD-associated genes that were initially linked to ASD because of their role at glutamatergic excitatory synapses was detected (Sauer et al., 2019). Among these were *Homer1* (homer protein homolog 1) (Kelleher et al., 2012), *Cask* (calcium/calmodulin-dependent serine protein kinase) (Iossifov et al., 2014; Stessman et al., 2017), *Cyfip1* (cytoplasmic FMR1 interacting protein 1) (Pathania et al., 2014), *Disc1* (disrupted In schizophrenia 1) (Kilpinen et al., 2008), *Dlg1* (discs, large homolog 1) (Li et al., 2014), *Fmr1* (fragile X mental retardation 1) (Rogers et al., 2001), *Mapk3* (mitogen-activated protein kinase 1 and 3/ERK1) (Schaaf et al., 2011), *Nf1* (neurofibromin 1) (Marui et al., 2004), and *Syngap1* (synaptic Ras GTPase activating protein 1) (Hamdan et al., 2011). Given that the encoded proteins are known to occur in a complex at synapses, it is likely that they also interact in GI epithelial cells and that the ASD-associated proteins have an additional role outside the CNS in the GI system.

Important roles of SHANK proteins in the GI tract were discovered. SHANK2 was reported to regulate fluid and ion transport in the intestine (Jung et al., 2014). Besides, SHANK3 may play a role in intestinal barrier formation, as an abnormal expression of tight junction proteins and increased barrier permeability were detected in *Shank3* knockout mice (Sauer et al., 2017). Thus, if *SHANK2* and *SHANK3* are regulated by zinc in the GI tract, which awaits confirmation, zinc deficiency may contribute to abnormal GI development and secretory responses to pathogens. A modification of GI morphology by altered zinc availability was already reported in pigs (Payne et al., 2006).

In addition, *SHANK3* expression in enterocytes differentiated from iPSCs regulates the expression of the zinc transporters *ZIP2* (*SLC39A2*) and *ZIP4* (*SLC39A2*) on mRNA and protein level. A loss of functional SHANK3 leads to reduced levels, especially of ZIP4 at the plasma membrane (Pfaender et al., 2017). ZIP4 is responsible for dietary zinc uptake into enterocytes at the apical membrane (Wang and Zhou, 2010). Mutations in ZIP4 have been associated with *Acrodermatitis enteropathica*, an inherited human zinc deficiency (Küry et al., 2002). To control zinc uptake, the expression of *Zip4* was shown to be highly responsive to dietary zinc in mice. Under zinc deficiency, *Zip4* was upregulated. Under increased zinc concentration, *Zip4* expression was downregulation, and the ZIP4 proteins were internalized and degraded (Dufner-Beattie et al., 2003). Knockdown and loss of *SHANK3* affect this regulatory mechanism of ZIP4 (Pfaender et al., 2017).

The functional interaction of SHANK3 and ZIP2 and ZIP4 was also confirmed in the GI system of *Shank3* knockout mice (Pfaender et al., 2017). Therefore the iPSC-based in vitro experiments reveal another link between genetic risk factors for ASD and trace metal homeostasis, which may contribute to the increased incidence rate of trace metal abnormalities beyond the pre- and perinatal period in many individuals with ASD (Fig. 11.1). While during healthy GI development, ASD-associated proteins that are also found in the GI system such as SHANK3 regulate intestinal morphology and barrier tightness, their loss, or dysregulation due to zinc

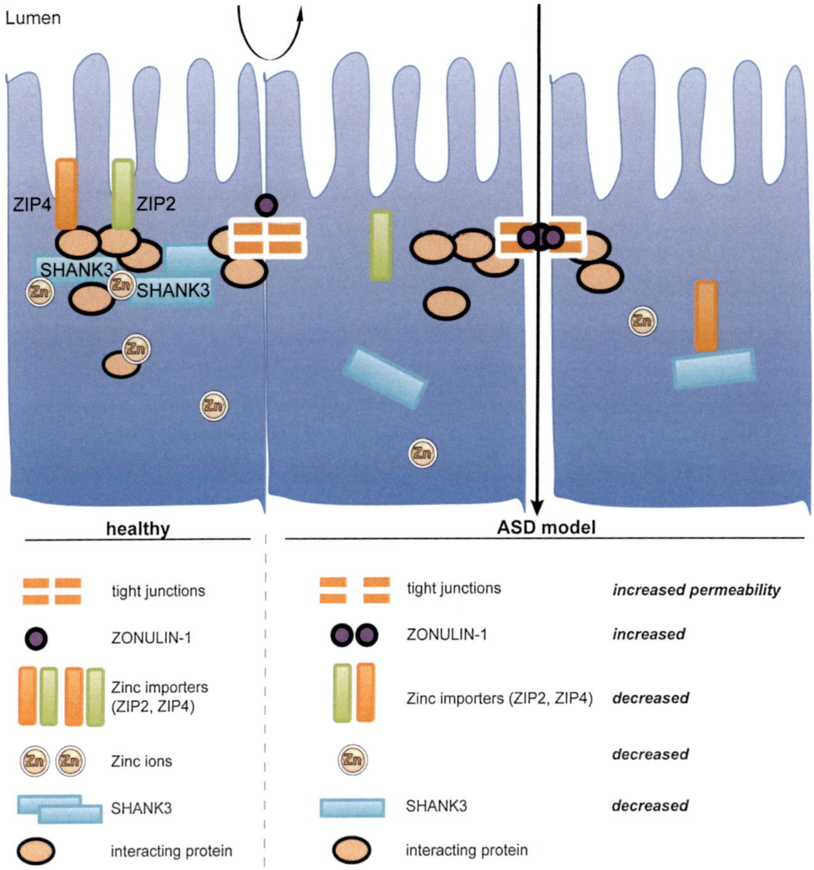

Figure 11.1 Left: In a healthy individual, SHANK3 protein complexes and interacting proteins, among them known ASD-associated genes, regulate GI morphology and barrier tightness through modification of ZONULIN-1 levels (Sauer et al., 2019). In addition, SHANK3 complexes regulate expression and localization of the zinc importers ZIP2 and ZIP4 at the apical membrane of enterocytes in the GI tract and mediate their regulatory responses to altered dietary zinc intake. Right: In ASD, the levels of SHANK3 or important interacting proteins decrease, or the proteins are dysfunctional due to mutations. This may result in increased barrier permeability (leaky gut) and low zinc intake (zinc deficiency). Both leaky gut and zinc deficiency are frequently reported in individuals with ASD.

deficiency, may result in abnormal GI physiology such as increased barrier permeability. Besides, mutations in SHANK3 proteins that are in complex with the zinc importers ZIP2 and ZIP4, or the loss of SHANK3 and interacting proteins, may lead to impaired zinc uptake into enterocytes. Impaired zinc uptake, in turn, will lead to low zinc status that will affect brain development and function and that may explain the low zinc status of many individuals with ASD. This process may be one mechanism of how gut–brain signaling occurs in ASD (see Chapter 12).

Manipulation of zinc levels in ASD

SHANK3 proteins were in a complex with ZIP2 and ZIP4 zinc importers in human iPSC-derived enterocytes and enterocytes of mice. In both, a loss of SHANK3 resulted in decreased ZIP4 expression and cell surface localization. Although this may be an explanation for the high incidence rates of zinc deficiency in PMDS (Grabrucker et al., 2014; Pfaender et al., 2017), it also highlights that individuals with ASD, at least those with mutations in SHANK3, but probably those with mutations in SHANK3 interaction partners as well, are facing challenges in response to low dietary zinc intake. Low dietary zinc intake results typically in a high expression of ZIP4 and high presence of the transporter at the cell surface. However, iPSC-derived enterocytes from PMDS patients did not show a significant upregulation of ZIP2 and ZIP4 in response to zinc depletion, while enterocytes from healthy individuals did (Pfaender et al., 2017). Therefore, individuals with PMDS may already be struggling to maintain an adequate zinc status when healthy individuals can still cope with a low zinc diet due to the upregulation of zinc importers, which maximizes zinc-uptake efficacy (Hagmeyer et al., 2018).

Based on this, it will be necessary for the future to discuss if the current recommended daily allowances/adequate intakes, and daily values (see Chapter 6) for zinc that are based on a 20%–30% absorption rate of zinc from dietary sources can be applied to individuals with ASD or specific subpopulations with ASD.

References

Dufner-Beattie, J., Wang, F., Kuo, Y.M., Gitschier, J., Eide, D., Andrews, G.K., 2003. The acrodermatitis enteropathica gene ZIP4 encodes a tissue-specific, zinc-regulated zinc transporter in mice. J Biol Chem 278 (35), 33474–33481.

Grabrucker, S., Jannetti, L., Eckert, M., Gaub, S., Chhabra, R., Pfaender, S., Mangus, K., Reddy, P.P., Rankovic, V., Schmeisser, M.J., Kreutz, M.R., Ehret, G., Boeckers, T.M., Grabrucker, A.M., 2014. Zinc deficiency dysregulates the synaptic ProSAP/Shank scaffold and might contribute to autism spectrum disorders. Brain 137 (Pt 1), 137–152.

Hagmeyer, S., Sauer, A.K., Grabrucker, A.M., 2018. Prospects of zinc supplementation in autism spectrum disorders and Shankopathies such as Phelan McDermid syndrome. Front Synaptic Neurosci 10, 11.

Hamdan, F.F., Daoud, H., Piton, A., Gauthier, J., Dobrzeniecka, S., Krebs, M.O., Joober, R., Lacaille, J.C., Nadeau, A., Milunsky, J.M., Wang, Z., Carmant, L., Mottron, L., Beauchamp, M.H., Rouleau, G.A., Michaud, J.L., 2011. De novo SYNGAP1 mutations in nonsyndromic intellectual disability and autism. Biol Psychiatry 69 (9), 898–901.

Horvath, K., Perman, J.A., 2002. Autism and gastrointestinal symptoms. Curr Gastroenterol Rep 4 (3), 251–258.

Iossifov, I., O'Roak, B.J., Sanders, S.J., Ronemus, M., Krumm, N., Levy, D., Stessman, H.A., Witherspoon, K.T., Vives, L., Patterson, K.E., Smith, J.D., Paeper, B., Nickerson, D.A., Dea, J., Dong, S., Gonzalez, L.E., Mandell, J.D., Mane, S.M., Murtha, M.T., Sullivan, C.A., Walker, M.F., Waqar, Z., Wei, L., Willsey, A.J., Yamrom, B., Lee, Y.H., Grabowska, E., Dalkic, E., Wang, Z., Marks, S., Andrews, P., Leotta, A., Kendall, J., Hakker, I., Rosenbaum, J., Ma, B., Rodgers, L., Troge, J., Narzisi, G., Yoon, S., Schatz, M.C., Ye, K., McCombie, W.R., Shendure, J., Eichler, E.E., State, M.W., Wigler, M., 2014. The contribution of de novo coding mutations to autism spectrum disorder. Nature 515 (7526), 216–221.

Jung, E.S., Park, J., Gee, H.Y., Jung, J., Noh, S.H., Lee, J.S., Richter, W., Namkung, W., Lee, M.G., 2014. Shank2 mutant mice display a hypersecretory response to cholera toxin. J Physiol 592 (8), 1809–1821.

Kelleher, R.J., Geigenmüller, U., Hovhannisyan, H., Trautman, E., Pinard, R., Rathmell, B., Carpenter, R., Margulies, D., 2012. High-throughput sequencing of mGluR signaling pathway genes reveals enrichment of rare variants in autism. PLoS One 7 (4), e35003.

Kilpinen, H., Ylisaukko-Oja, T., Hennah, W., Palo, O.M., Varilo, T., Vanhala, R., Nieminen-von Wendt, T., von Wendt, L., Paunio, T., Peltonen, L., 2008. Association of DISC1 with autism and Asperger syndrome. Mol Psychiatry 13 (2), 187–196.

Kolevzon, A., Angarita, B., Bush, L., Wang, A.T., Frank, Y., Yang, A., Rapaport, R., Saland, J., Srivastava, S., Farrell, C., Edelmann, L.J., Buxbaum, J.D., 2014. Phelan-McDermid syndrome: a review of the literature and practice parameters for medical assessment and monitoring. J Neurodev Disord 6 (1), 39.

Krug, A.K., Kolde, R., Gaspar, J.A., Rempel, E., Balmer, N.V., Meganathan, K., Vojnits, K., Baquié, M., Waldmann, T., Ensenat-Waser, R., Jagtap, S., Evans, R.M., Julien, S., Peterson, H., Zagoura, D., Kadereit, S., Gerhard, D., Sotiriadou, I., Heke, M., Natarajan, K., Henry, M., Winkler, J., Marchan, R., Stoppini, L., Bosgra, S., Westerhout, J., Verwei, M., Vilo, J., Kortenkamp, A., Hescheler, J., Hothorn, L., Bremer, S., van Thriel, C., Krause, K.H., Hengstler, J.G., Rahnenführer, J., Leist, M., Sachinidis, A., 2013. Human embryonic stem cell-derived test systems for developmental neurotoxicity: a transcriptomics approach. Arch Toxicol 87 (1), 123–143.

Küry, S., Dréno, B., Bézieau, S., Giraudet, S., Kharfi, M., Kamoun, R., Moisan, J.P., 2002. Identification of SLC39A4, a gene involved in acrodermatitis enteropathica. Nat Genet 31 (3), 239–240.

Leblond, C.S., Nava, C., Polge, A., Gauthier, J., Huguet, G., Lumbroso, S., Giuliano, F., Stordeur, C., Depienne, C., Mouzat, K., Pinto, D., Howe, J., Lemière, N., Durand, C.M., Guibert, J., Ey, E., Toro, R., Peyre, H., Mathieu, A., Amsellem, F., Rastam, M., Gillberg, I.C., Rappold, G.A., Holt, R., Monaco, A.P., Maestrini, E., Galan, P., Heron, D., Jacquette, A., Afenjar, A., Rastetter, A., Brice, A., Devillard, F., Assouline, B., Laffargue, F., Lespinasse, J., Chiesa, J., Rivier, F., Bonneau, D., Regnault, B., Zelenika, D., Delepine, M., Lathrop, M., Sanlaville, D., Schluth-Bolard, C., Edery, P., Perrin, L., Tabet, A.C., Schmeisser, M.J., Boeckers, T.M., Coleman, M., Sato, D., Szatmari, P., Scherer, S.W., Rouleau, G.A., Betancur, C., Leboyer, M., Gillberg, C., Delorme, R., Bourgeron, T., 2014. Meta-analysis of

SHANK mutations in autism spectrum disorders: a gradient of severity in cognitive impairments. PLoS Genet 10 (9), e1004580.
Li, J., Shi, M., Ma, Z., Zhao, S., Euskirchen, G., Ziskin, J., Urban, A., Hallmayer, J., Snyder, M., 2014. Integrated systems analysis reveals a molecular network underlying autism spectrum disorders. Mol Syst Biol 10, 774.
Marui, T., Hashimoto, O., Nanba, E., Kato, C., Tochigi, M., Umekage, T., Ishijima, M., Kohda, K., Kato, N., Sasaki, T., 2004. Association between the neurofibromatosis-1 (NF1) locus and autism in the Japanese population. Am J Med Genet B Neuropsychiatr Genet 131B (1), 43–47.
O'Roak, B.J., Deriziotis, P., Lee, C., Vives, L., Schwartz, J.J., Girirajan, S., Karakoc, E., Mackenzie, A.P., Ng, S.B., Baker, C., Rieder, M.J., Nickerson, D.A., Bernier, R., Fisher, S.E., Shendure, J., Eichler, E.E., 2011. Exome sequencing in sporadic autism spectrum disorders identifies severe de novo mutations. Nat Genet 43 (6), 585–589.
O'Roak, B.J., Vives, L., Girirajan, S., Karakoc, E., Krumm, N., Coe, B.P., Levy, R., Ko, A., Lee, C., Smith, J.D., Turner, E.H., Stanaway, I.B., Vernot, B., Malig, M., Baker, C., Reilly, B., Akey, J.M., Borenstein, E., Rieder, M.J., Nickerson, D.A., Bernier, R., Shendure, J., Eichler, E.E., 2012. Sporadic autism exomes reveal a highly interconnected protein network of de novo mutations. Nature 485 (7397), 246–250.
Park, I.H., Arora, N., Huo, H., Maherali, N., Ahfeldt, T., Shimamura, A., Lensch, M.W., Cowan, C., Hochedlinger, K., Daley, G.Q., 2008. Disease-specific induced pluripotent stem cells. Cell 134 (5), 877–886.
Pathania, M., Davenport, E.C., Muir, J., Sheehan, D.F., López-Doménech, G., Kittler, J.T., 2014. The autism and schizophrenia associated gene CYFIP1 is critical for the maintenance of dendritic complexity and the stabilization of mature spines. Transl Psychiatry 4, e374.
Payne, R.L., Bidner, T.D., Fakler, T.M., Southern, L.L., 2006. Growth and intestinal morphology of pigs from sows fed two zinc sources during gestation and lactation. J Anim Sci 84 (8), 2141–2149.
Pfaender, S., Föhr, K., Lutz, A.K., Putz, S., Achberger, K., Linta, L., Liebau, S., Boeckers, T.M., Grabrucker, A.M., 2016. Cellular zinc homeostasis contributes to neuronal differentiation in human induced pluripotent stem cells. Neural Plast 2016, 3760702.
Pfaender, S., Sauer, A.K., Hagmeyer, S., Mangus, K., Linta, L., Liebau, S., Bockmann, J., Huguet, G., Bourgeron, T., Boeckers, T.M., Grabrucker, A.M., 2017. Zinc deficiency and low enterocyte zinc transporter expression in human patients with autism related mutations in SHANK3. Sci Rep 7, 45190.
Rogers, S.J., Wehner, D.E., Hagerman, R., 2001. The behavioral phenotype in fragile X: symptoms of autism in very young children with fragile X syndrome, idiopathic autism, and other developmental disorders. J Dev Behav Pediatr 22 (6), 409–417.
Russo, F.B., Brito, A., de Freitas, A.M., Castanha, A., de Freitas, B.C., Beltrão-Braga, P.C.B., 2019. The use of iPSC technology for modeling Autism Spectrum Disorders. Neurobiol Dis 130, 104483.
Sauer, A.K., Pfaender, S., Hagmeyer, S., Tarana, L., Mattes, A.K., Briel, F., Küry, S., Boeckers, T.M., Grabrucker, A.M., 2017. Characterization of zinc amino acid complexes for zinc delivery in vitro using Caco-2 cells and enterocytes from hiPSC. Biometals 30 (5), 643–661.
Sauer, A.K., Bockmann, J., Steinestel, K., Boeckers, T.M., Grabrucker, A.M., 2019. Altered intestinal morphology and Microbiota composition in the autism spectrum disorders associated SHANK3 mouse model. Int J Mol Sci 20 (9).
Schaaf, C.P., Sabo, A., Sakai, Y., Crosby, J., Muzny, D., Hawes, A., Lewis, L., Akbar, H., Varghese, R., Boerwinkle, E., Gibbs, R.A., Zoghbi, H.Y., 2011. Oligogenic

heterozygosity in individuals with high-functioning autism spectrum disorders. Hum Mol Genet 20 (17), 3366–3375.
Stessman, H.A., Xiong, B., Coe, B.P., Wang, T., Hoekzema, K., Fenckova, M., Kvarnung, M., Gerdts, J., Trinh, S., Cosemans, N., Vives, L., Lin, J., Turner, T.N., Santen, G., Ruivenkamp, C., Kriek, M., van Haeringen, A., Aten, E., Friend, K., Liebelt, J., Barnett, C., Haan, E., Shaw, M., Gecz, J., Anderlid, B.M., Nordgren, A., Lindstrand, A., Schwartz, C., Kooy, R.F., Vandeweyer, G., Helsmoortel, C., Romano, C., Alberti, A., Vinci, M., Avola, E., Giusto, S., Courchesne, E., Pramparo, T., Pierce, K., Nalabolu, S., Amaral, D.G., Scheffer, I.E., Delatycki, M.B., Lockhart, P.J., Hormozdiari, F., Harich, B., Castells-Nobau, A., Xia, K., Peeters, H., Nordenskjöld, M., Schenck, A., Bernier, R.A., Eichler, E.E., 2017. Targeted sequencing identifies 91 neurodevelopmental-disorder risk genes with autism and developmental-disability biases. Nat Genet 49 (4), 515–526.
Takahashi, K., Yamanaka, S., 2006. Induction of pluripotent stem cells from mouse embryonic and adult fibroblast cultures by defined factors. Cell 126 (4), 663–676.
Wang, X., Zhou, B., 2010. Dietary zinc absorption: a play of Zips and ZnTs in the gut. IUBMB Life 62 (3), 176–182.

CHAPTER 12

Extracerebral biometals in autism spectrum disorders: the gut–brain axis

Contents

Recently, pieces of evidence have been presented in a vastly growing field of research into gut–brain interactions, which indicate that pathology of the gastrointestinal (GI) system during development may play a causative or modifying role in ASD (Saurman et al., 2020). GI disturbances are frequent in ASD (Lefter et al., 2019). Several studies link behavioral difficulties in ASD to GI problems in humans (Horvath and Perman, 2002; Nikolov et al., 2009). For example, individuals with ASD frequently have symptoms that are associated with an abnormal intestinal barrier function (de Magistris et al., 2010; Vela et al., 2015). Besides, alterations in the composition of microbiota have been consistently reported in individuals with ASD and ASD mouse models that lead to alterations in the levels of certain bacterial species, and the presence or absence of specific species in the gut microbiome (Finegold et al., 2002; Parracho et al., 2005). It has been speculated that altered microbiota profiles may be both cause and consequence of ASD. Interestingly, the severity of GI disturbances was found to correlate with the severity of ASD behavioral pathology in children with ASD (Adams et al., 2011).

There are several mechanisms discussed how an altered microbiome can affect brain development and function. In brief, gut–brain communication may occur through endocrine messaging via immune cells localized in the gut and the release of cytokines, through the enteric nervous system and its neurons that are connected to the brain via the vagus nerve, and through bacterial compounds and metabolites that may influence brain cells.

Biometals in Autism Spectrum Disorders
ISBN 978-0-12-821132-8
https://doi.org/10.1016/B978-0-12-821132-8.00012-4

Especially, the release of cytokines and the presence of bacterial neuroactive molecules may have relevance for trace metal biology.

The release of proinflammatory cytokines may occur as an immune response to GI system alterations, such as increased intestinal barrier permeability with the consequential passing of bacteria or bacterial compounds into the host circulation. The proinflammatory processes may then ultimately affect brain development through the presence of cytokines such as IL-1β, IL-6, IL-18, and TNF-α (Dinan and Cryan, 2017; Siniscalco et al., 2018), bacterial substances, or, as discussed in Chapter 10, through altered zinc levels due to immune activation. Chronic inflammation, especially within the gut, can damage the epithelial cell layer and further increase intestinal barrier permeability (de Theije et al., 2011).

Altered microbiome profiles may because and consequence of increased intestinal permeability, and also influence inflammatory processes. In line with human studies, systemic inflammation and subsequent neuroinflammation were also observed in animal models of ASD with altered microbiota profiles (Siniscalco et al., 2018; Needham et al., 2018).

In humans, the microbiome is established at birth. Very shortly following vaginal birth, the microbiome composition of an infant resembles that of the mother's vagina (Rodriguez et al., 2015). It will then change over time, until from early childhood on, the microbiota composition of the GI system is relatively stable (Gilbert et al., 2018; Rodríguez et al., 2015). Several factors can influence the gut microbiota composition. For example, if the birth occurred by C-Section, the baby is not exposed to the vaginal microbiome of the mother, which influences the first microbiome that is established. Later, the use of antibiotics, nutrition, environmental stressors, and the host genetics will influence the microbiota composition of an individual (Dong and Gupta, 2019; Hasan and Yang, 2019; Rodriguez et al., 2015). One factor that influences trace metal composition is the exposure to toxic metals and bioavailability of trace metals. As is the host, the microbiota species in the GI system are dependent on essential trace metals such as zinc. For example, in rats, it was estimated that about 20% of the dietary intake of zinc is used by intestinal bacteria (Smith et al., 1972). However, some species may have more effective mechanisms of zinc uptake and/or may be less dependent on zinc. These species will strive under low-zinc conditions and outcompete other bacteria. Thus, their abundance will increase, while the abundance of bacteria that are less competitive for zinc will decrease. The same mechanisms may occur for other trace metals, including toxic trace metals, where some bacteria might have a high tolerance, but others will be affected.

Microbiome alterations as a result of trace metal abnormalities

In healthy adults, bacteria belonging to the phyla Firmicutes and Bacteroidetes are prominent. However, their relative abundance and the abundance of different families and certain species within them can vary (Eckburg et al., 2005; Dinan and Cryan, 2017). Both phyla are also abundant in rodent gut microbiomes. The exposure to toxic metals has effects on the gut microbiota composition. For example, acute methylmercury overload induces alterations in the abundance of specific gut microbiota in rats. On the phylum level, a decrease in Bacteriodetes and Proteobacteria was seen, and an increase in Firmicutes. On the family level, Lactobacillaceae, Bacteroidaceae, Streptococcaceae, and Sutterellaceae decreased, while Desulfovibrionaceae, Helicobacteraceae, Peptococcaceae, and Rhodospirillaceae increased (Lin et al., 2020).

Furthermore, acute exposure to cadmium causes a lower microbial diversity in the gut of mice. A decrease in Bacteriodetes and an increase in Proteobacteria was seen, among other changes (Li et al., 2019). In rats, cadmium seems to increase the genus Akkermansia (phylum Verrucomicrobia), Allobaculum (phylum Firmicutes), Bifidobacterium (phylum Actinobacteria), and to decrease the abundance of Ruminococcus (phylum Firmicutes) (Richardson et al., 2018).

In addition, acute and perinatal lead exposure induce gut microbiota alterations (Zhai et al., 2019; Wu et al., 2016). In mice, after perinatal lead overload, the abundance of bacteria from the phyla Bacteroidetes was decreased and from Firmicutes was increased with lead exposure, and the whole bacterial composition was significantly different (Wu et al., 2016) with more anaerobe species in lead-exposed mice.

In pregnant women, it was found that changes in the genus Subdoligranulum (phylum Firmicutes, class: Clostridia) are significantly associated with biomarkers for methyl-mercury exposure levels (and stool methylmercury, stool inorganic mercury, or hair total mercury) (Rotherberg et al., 2016). Further, in humans, the abundance of the family Lachnospiraceae (phylum Firmicutes, class Clostridia) and Erysipelotrichaceae (phylum Firmicutes, class Erysipelotrichia), which are linked to intestinal inflammation, were found increased in individuals exposed to heavy metals (Shao and Zhu, 2020).

However, microbiota species do not only change in response to toxic trace metals, but they also influence the levels and exposure to them. For example, in rats, antibiotics were shown to alter the microbiota composition in the GI system in a way that prevents normal gut anaerobes from converting methyl-mercury to inorganic mercury that can be more easily excreted (Adams et al., 2007). Thus, the microbiota composition of the gut may either facilitate the excretion of mercury or contribute to its increase in the body. This regulation may similarly occur for other trace metals.

In addition to toxic metals, essential metals influence the microbiome. For example, acute zinc deficiency was reported to affect microbiota composition in rodents and chicken (Reed et al., 2015, 2018). In particular, a decrease in Veruccomicrobia and an increase in Firmicutes has been observed (Mayneris-Perxachs et al., 2016; Lopez and Skaar, 2018). As ASDs are neurodevelopmental disorders that may be caused by pathologies that are already present in utero, an altered microbiota composition may be inherited from the mother that is exposed to toxic metals or deficient in essential trace metals, rather than acquired independently later in life. Following this hypothesis, the gut microbiome of zinc-deficient mice during pregnancy was analyzed, whose offspring shows ASD-like behavior. The results show that zinc deficiency during pregnancy leads to a significant alteration in microbiota composition in the gut of the mother (Sauer and Grabrucker, 2019).

In line with the results of nonpregnant animals, pregnant zinc-deficient mice showed an increase in the phylum Firmicutes. In addition, an increase in Actinobacteria and Bacteroidetes, and a decrease in Proteobacteria was observed (Sauer and Grabrucker, 2019). Actinobacteria are generally regarded as so-called pathobionts that, under certain circumstances, can promote disease. In particular, Actinobacteria are associated with chronic inflammatory conditions such as inflammatory bowel disease (Frank et al., 2007; Morgan et al., 2012).

The increase in Firmicutes is attributed to higher levels of Clostridium, Oxobacter, Lachnoclostridium, Marvinbryantia, and Murimonas found in the microbiome of pregnant zinc-deficient mice. The increase in Actinobacteria is due to an increase in Bifidobacterium, Olsenella, Parvibacter, Asaccharobacter, Eggerthella, and Enterorhabdus genera. The increase in Bacteroidetes can be attributed to an increase in Odoribacter and Alistripes at a genus level. The decrease of Proteobacteria is due to a loss of Kiloniella, Rhodospirilium, Lautropia, and Helicobacter at a genus level, and finally, the decrease that was seen in the phylum Verrucomicrobia is solely due to a decrease in the genus Akkermansia. The results on the level of bacteria families are shown in Table 12.1.

Table 12.1 Results for families of bacteria upregulated/downregulated between zinc-deficient pregnant mice and their controls.

Taxonomy (phylum, class, order, family)	Regulation (in Zn def. mice)	Significance
Actinobacteria, Actinobacteria, Bifidobacteriales, Bifidobacteriaceae	▲	**
Actinobacteria, Actinobacteria, geodermatophilales, Geodermatophilaceae	↔	ns
Actinobacteria, Actinobacteria, Micrococcales, Micrococcaceae	↔	ns
Actinobacteria, Actinobacteria, Propionibacteriales, Propionibacteriaceae	↔	ns
Actinobacteria, Coriobacteriia, Coriobacteriales, Atopobiaceae	▲	***
Actinobacteria, Coriobacteriia, Coriobacteriales, Coriobacteriaceae	▲	**
Actinobacteria, Coriobacteriia, Eggerthellales, Eggerthellaceae	▲	***
Bacteroidetes, Bacteroidia, Bacteroidales, Porphyromonadaceae	▲	**
Bacteroidetes, Bacteroidia, Bacteroidales, Prevotellaceae	↔	ns
Bacteroidetes, Bacteroidia, Bacteroidales, Rikenellaceae	▲	***
Chlamydiae, Chlamydiia, Chlamydiales, Chlamydiaceae	↔	ns
Firmicutes, Bacilli, Lactobacillales, Enterococcaceae	▲	**
Firmicutes, Bacilli, Lactobacillales, Lactobacillaceae	▲	**
Firmicutes, Bacilli, Lactobacillales, Streptococcaceae	↔	ns
Firmicutes, Clostridia, Clostridiales, Catabacteriaceae	↔	ns
Firmicutes, Clostridia, Clostridiales, Christensenellaceae	↔	ns
Firmicutes, Clostridia, Clostridiales, Clostridiaceae	▲	**
Firmicutes, Clostridia, Clostridiales, Clostridiales Family XIII. Incertae Sedis	▲	**
Firmicutes, Clostridia, Clostridiales, Eubacteriaceae	▲	**
Firmicutes, Clostridia, Clostridiales, Gracilibacteraceae	↔	ns
Firmicutes, Clostridia, Clostridiales, Lachnospiraceae	▲	***
Firmicutes, Clostridia, Clostridiales, Oscillospiraceae	▲	*
Firmicutes, Clostridia, Clostridiales, Peptococcaceae	▲	**
Firmicutes, Clostridia, Clostridiales, Proteinivoraceae	↔	ns

Continued

Table 12.1 Results for families of bacteria upregulated/downregulated between zinc-deficient pregnant mice and their controls.—cont'd

Taxonomy (phylum, class, order, family)	Regulation (in Zn def. mice)	Significance
Firmicutes, Clostridia, Clostridiales, Ruminococcaceae	▼	*
Firmicutes, Erysipelotrichia, Erysipelotrichales, Erysipelotrichaceae	▲	**
Nitrospinae, Nitrospinia, Nitrospinales, Nitrospinaceae	↔	ns
Proteobacteria, Alphaproteobacteria, Kiloniellales, Kiloniellaceae	▼	**
Proteobacteria, Alphaproteobacteria, Rhizobiales, Cohaesibacteraceae	↔	ns
Proteobacteria, Alphaproteobacteria, Rhizobiales, Hyphomicrobiaceae	↔	ns
Proteobacteria, Alphaproteobacteria, Rhizobiales, Methylobacteriaceae	↔	ns
Proteobacteria, Alphaproteobacteria, Rhodobacterales, Hyphomonadaceae	↔	ns
Proteobacteria, Alphaproteobacteria, Rhodospirillales, Rhodospirillaceae	▼	**
Proteobacteria, Betaproteobacteria, Burkholderiales, Burkholderiaceae	▼	*
Proteobacteria, Betaproteobacteria, Burkholderiales, Comamonadaceae	▼	**
Proteobacteria, Betaproteobacteria, Burkholderiales, Sutterellaceae	↔	ns
Proteobacteria, Deltaproteobacteria, Desulfovibrionales, Desulfovibrionaceae	▲	*
Proteobacteria, Epsilonproteobacteria, Campylobacterales, Helicobacteraceae	▼	**
Proteobacteria, gammaproteobacteria, Enterobacteriales, Enterobacteriaceae	↔	ns
Tenericutes, Mollicutes, Mycoplasmatales, Mycoplasmataceae	↔	ns
Verrucomicrobia, Verrucomicrobiae, Verrucomicrobiales, Akkermansiaceae	▼	**

The various families of bacteria are displayed and significant upregulation (*up arrow*), downregulation (*down arrow*), or no significant alteration (*side-to-side arrow*) indicated. * indicates a P value with $P \leq .05$, $^{**}P \leq .01$, and $^{***}P \leq .001$ (Grabrucker, original data).

There are similarities in the alterations of the microbiota composition of zinc-deficient pregnant mice to those reported from human and animal models for ASD. For example, alterations in the phylum Firmicutes and in particular bacteria of the genus Clostridium (phylum Firmicutes) were consistently reported upregulated in ASD (both human patients and murine ASD models) (Ding et al., 2017; Hughes et al., 2018; Rosenfeld, 2015; Vuong and Hsiao, 2017; Warner, 2019; Sauer et al., 2019) and are also altered in pregnant zinc-deficient mice. These changes recently emerged as one of the few characteristic abnormalities in a meta-analysis of ASD microbiome studies (Ho et al., 2020). Besides, bacteria of the genus Desulfovibrio (phylum Proteobacteria) were consistently reported upregulated in ASD human patients (Berding and Donovan, 2016; Hughes et al., 2018; Rosenfeld, 2015; Tomova et al., 2015) and the family of Desulfovibrionaceae is also increased in zinc-deficient mice.

Inflammation as a result of trace metal abnormalities

Several studies support the idea that intestinal barrier dysbiosis and altered microbiota profiles trigger systemic inflammatory responses that may ultimately result in neuroinflammation. A key proinflammatory cytokine in this process is Interleukin 6 (IL-6). IL-6 was already shown to be a key mediator of ASD behavior in mice with maternal immune activation (MIA) (Wei et al., 2013) that also have low zinc status. A recent study confirmed that zinc deficiency during pregnancy not only alters the maternal microbiome but also increases IL-6 levels (Sauer and Grabrucker, 2019). This interplay of zinc and IL-6 was also reported before in aging mice (Wong et al., 2015). Blocking placental IL-6 signaling was able to prevent behavioral abnormalities in MIA mice (Wu et al., 2017), as was zinc supplementation (Kirsten et al., 2015a,b). Thus, a close relationship exists between gastrointestinal disturbances, zinc deficiency, and IL-6 signaling.

Zinc plays an important role in regulating immune activation. Low zinc status elevates the inflammatory response through the modulation (activation) of nuclear factor kappa B (NF-κB) and by decreased control of oxidative stress (Gammoh and Rink, 2017). NF-κB activation induces IL-6 expression. Therefore, GI disturbances such as altered microbiota composition and increased intestinal barrier permeability trigger processes that may lower zinc availability, which in turn will facilitate proinflammatory signaling (e.g., via the release of IL-6).

Toxic metals were also shown to induce IL-6 signaling. For example, exposure to high levels of lead pre- and perinatally resulted in a significant increase in IL-6 (and IL-1β and TGF-β) in the brain of pups (Chibowska et al., 2020). The presence of lead increases the expression of IL-6 (Kishikawa and Lawrence, 1998) after immune activation. Besides, methylmercury affects the expression of proinflammatory cytokines, such as IL-6 (and IL-8) (Yamamoto et al., 2017). Thus, inflammatory gut–brain signaling caused by GI disturbances is exacerbated by the presence of toxic metals and the absence of zinc.

Conclusions

ASD is a heterogeneous group of disorders with core symptoms that may be present with different severity and a variety of comorbidities that may also be present in different combinations and different severity. This clinical picture fits well with the various roles of trace metals in the body that are not restricted to the brain. For example, while prenatal zinc deficiency has apparent effects on neurogenesis, synaptogenesis, and ultimately connectivity in the developing brain, additional extracerebral pathology develops that may interact with brain pathologies in a complicated way such as microbiota–gut–brain signaling. Therefore, it may not be possible to explain the causative role of trace metal imbalances for ASD by one single mechanism, but only by the combination of several, and probably yet unexplored, biological processes affected by these imbalances. Although this makes the science of biometals in ASD a complicated field of research, it may ultimately lead to a model that may explain the occurrence and interplay of all features seen in individuals with ASD.

Taken together, a model can be proposed for the mechanisms of zinc deficiency that play a role in ASD (Fig. 12.1): In parallel to causing synaptic deficits in the CNS of the developing offspring, zinc deficiency alters the microbiome of the pregnant mouse. The resulting abnormal microbiota composition leads to maternal immune activation, which promotes the release of proinflammatory cytokines. These cytokines may further aggravate maternal hypozincemia. At birth, the offspring inherits the altered microbiome from the mother, which may affect the GI physiology, and, together with effects triggered by maternal circulating cytokines, may facilitate a chronic inflammatory state. Thus, microbiota alterations and inflammation will additionally affect the developing brain through microbiota–gut–brain signaling and contribute to brain pathologies, thereby modifying the ASD phenotype and its severity.

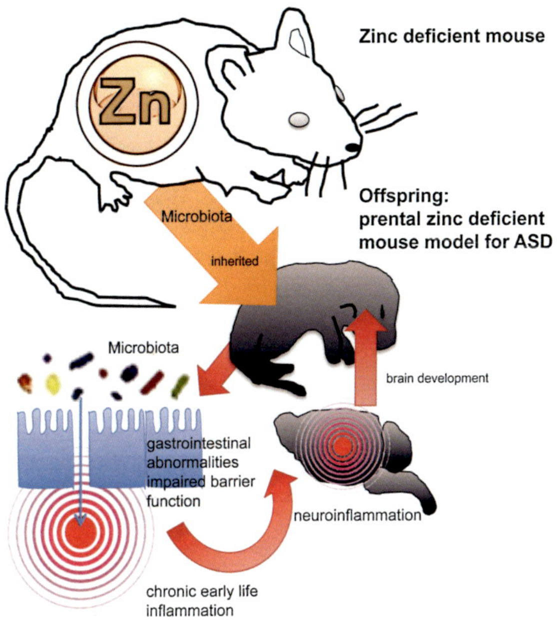

Figure 12.1 Zinc deficiency alters the microbiome of pregnant mice and causes GI disturbances that result in maternal immune activation. At birth, the offspring inherits the altered microbiome from the mother, which may affect its GI development and physiology resulting in a chronic early life inflammatory state. Inflammation with its associated release of cytokines will additionally affect the developing brain of prenatal zinc-deficient mice through microbiota—gut—brain signaling, and thereby modify the ASD phenotype and its severity.

Similarly, exposure to toxic metals may alter the microbiota composition, thereby causing immune activation and hypozincemia. This would explain the crosstalk of toxic metals beyond the competition for proteins with binding sites that have an affinity for essential as well as toxic metals. Hypozincemia will then, as discussed before, affect brain development and function.

References

Adams, J.B., Johansen, L.J., Powell, L.D., Quig, D., Rubin, R.A., 2011. Gastrointestinal flora and gastrointestinal status in children with autism—comparisons to typical children and correlation with autism severity. BMC Gastroenterol 11, 22.

Adams, J.B., Romdalvik, J., Ramanujam, V.M., Legator, M.S., 2007. Mercury, lead, and zinc in baby teeth of children with autism versus controls. J Toxicol Environ Health A 70 (12), 1046—1051.

Berding, K., Donovan, S.M., 2016. Microbiome and nutrition in autism spectrum disorder: current knowledge and research needs. Nutr Rev 74 (12), 723—736.

Chibowska, K., Korbecki, J., Gutowska, I., Metryka, E., Tarnowski, M., Goschorska, M., Barczak, K., Chlubek, D., Baranowska-Bosiacka, I., 2020. Pre- and neonatal exposure to lead (Pb) induces neuroinflammation in the forebrain cortex, hippocampus and cerebellum of rat pups. Int J Mol Sci 21 (3).

de Magistris, L., Familiari, V., Pascotto, A., Sapone, A., Frolli, A., Iardino, P., Carteni, M., De Rosa, M., Francavilla, R., Riegler, G., Militerni, R., Bravaccio, C., 2010. Alterations of the intestinal barrier in patients with autism spectrum disorders and in their first-degree relatives. J Pediatr Gastroenterol Nutr 51 (4), 418–424.

de Theije, C.G., Wu, J., da Silva, S.L., Kamphuis, P.J., Garssen, J., Korte, S.M., Kraneveld, A.D., 2011. Pathways underlying the gut-to-brain connection in autism spectrum disorders as future targets for disease management. Eur J Pharmacol 668 (Suppl. 1), 70–80.

Dinan, T.G., Cryan, J.F., 2017. Brain-gut-microbiota Axis and mental health. Psychosom Med 79 (8), 920–926.

Ding, H.T., Taur, Y., Walkup, J.T., 2017. Gut microbiota and autism: key concepts and findings. J Autism Dev Disord 47 (2), 480–489.

Dong, T.S., Gupta, A., 2019. Influence of early life, diet, and the environment on the microbiome. Clin Gastroenterol Hepatol 17 (2), 231–242.

Eckburg, P.B., Bik, E.M., Bernstein, C.N., Purdom, E., Dethlefsen, L., Sargent, M., Gill, S.R., Nelson, K.E., Relman, D.A., 2005. Diversity of the human intestinal microbial flora. Science 308 (5728), 1635–1638.

Finegold, S.M., Molitoris, D., Song, Y., Liu, C., Vaisanen, M.L., Bolte, E., McTeague, M., Sandler, R., Wexler, H., Marlowe, E.M., Collins, M.D., Lawson, P.A., Summanen, P., Baysallar, M., Tomzynski, T.J., Read, E., Johnson, E., Rolfe, R., Nasir, P., Shah, H., Haake, D.A., Manning, P., Kaul, A., 2002. Gastrointestinal microflora studies in late-onset autism. Clin Infect Dis 35 (Suppl. 1), 6–16.

Frank, D.N., St Amand, A.L., Feldman, R.A., Boedeker, E.C., Harpaz, N., Pace, N.R., 2007. Molecular-phylogenetic characterization of microbial community imbalances in human inflammatory bowel diseases. Proc Natl Acad Sci USA 104 (34), 13780–13785.

Gammoh, N.Z., Rink, L., 2017. Zinc in infection and inflammation. Nutrients 9 (6).

Gilbert, J.A., Blaser, M.J., Caporaso, J.G., Jansson, J.K., Lynch, S.V., Knight, R., 2018. Current understanding of the human microbiome. Nat Med 24 (4), 392–400.

Hasan, N., Yang, H., 2019. Factors affecting the composition of the gut microbiota, and its modulation. PeerJ 7, e7502.

Ho, L.K.H., Tong, V.J.W., Syn, N., Nagarajan, N., Tham, E.H., Tay, S.K., Shorey, S., Tambyah, P.A., Law, E.C.N., 2020. Gut microbiota changes in children with autism spectrum disorder: a systematic review. Gut Pathog 12, 6.

Horvath, K., Perman, J.A., 2002. Autistic disorder and gastrointestinal disease. Curr Opin Pediatr 14 (5), 583–587.

Hughes, H.K., Rose, D., Ashwood, P., 2018. The gut microbiota and dysbiosis in autism spectrum disorders. Curr Neurol Neurosci Rep 18 (11), 81.

Kirsten, T.B., Chaves-Kirsten, G.P., Bernardes, S., Scavone, C., Sarkis, J.E., Bernardi, M.M., Felicio, L.F., 2015a. Lipopolysaccharide exposure induces maternal hypozincemia, and prenatal zinc treatment prevents autistic-like behaviors and disturbances in the striatal dopaminergic and mTOR systems of offspring. PLoS One 10 (7), e0134565.

Kirsten, T.B., Queiroz-Hazarbassanov, N., Bernardi, M.M., Felicio, L.F., 2015b. Prenatal zinc prevents communication impairments and BDNF disturbance in a rat model of autism induced by prenatal lipopolysaccharide exposure. Life Sci 130, 12–17.

Kishikawa, H., Lawrence, D.A., 1998. Differential production of interleukin-6 in the brain and spleen of mice treated with lipopolysaccharide in the presence and absence of lead. J Toxicol Environ Health A 53 (5), 357–373.

Lefter, R., Ciobica, A., Timofte, D., Stanciu, C., Trifan, A., 2019. A descriptive review on the prevalence of gastrointestinal disturbances and their multiple associations in autism spectrum disorder. Medicina (Kaunas) 56 (1).

Li, X., Brejnrod, A.D., Ernst, M., Rykær, M., Herschend, J., Olsen, N.M.C., Dorrestein, P.C., Rensing, C., Sørensen, S.J., 2019. Heavy metal exposure causes changes in the metabolic health-associated gut microbiome and metabolites. Environ Int 126, 454–467.

Lin, X., Zhao, J., Zhang, W., He, L., Wang, L., Chang, D., Cui, L., Gao, Y., Li, B., Chen, C., Li, Y.F., 2020. Acute oral methylmercury exposure perturbs the gut microbiome and alters gut-brain axis related metabolites in rats. Ecotoxicol Environ Saf 190, 110130.

Lopez, C.A., Skaar, E.P., 2018. The impact of dietary transition metals on host-bacterial interactions. Cell Host Microbe 23 (6), 737–748.

Mayneris-Perxachs, J., Bolick, D.T., Leng, J., Medlock, G.L., Kolling, G.L., Papin, J.A., Swann, J.R., Guerrant, R.L., 2016. Protein- and zinc-deficient diets modulate the murine microbiome and metabolic phenotype. Am J Clin Nutr 104 (5), 1253–1262.

Morgan, X.C., Tickle, T.L., Sokol, H., Gevers, D., Devaney, K.L., Ward, D.V., Reyes, J.A., Shah, S.A., LeLeiko, N., Snapper, S.B., Bousvaros, A., Korzenik, J., Sands, B.E., Xavier, R.J., Huttenhower, C., 2012. Dysfunction of the intestinal microbiome in inflammatory bowel disease and treatment. Genome Biol 13 (9), R79.

Needham, B.D., Tang, W., Wu, W.L., 2018. Searching for the gut microbial contributing factors to social behavior in rodent models of autism spectrum disorder. Dev Neurobiol 78 (5), 474–499.

Nikolov, R.N., Bearss, K.E., Lettinga, J., Erickson, C., Rodowski, M., Aman, M.G., McCracken, J.T., McDougle, C.J., Tierney, E., Vitiello, B., Arnold, L.E., Shah, B., Posey, D.J., Ritz, L., Scahill, L., 2009. Gastrointestinal symptoms in a sample of children with pervasive developmental disorders. J Autism Dev Disord 39 (3), 405–413.

Parracho, H.M., Bingham, M.O., Gibson, G.R., McCartney, A.L., 2005. Differences between the gut microflora of children with autistic spectrum disorders and that of healthy children. J Med Microbiol 54 (Pt 10), 987–991.

Reed, S., Knez, M., Uzan, A., Stangoulis, J.C.R., Glahn, R.P., Koren, O., Tako, E., 2018. Alterations in the gut (Gallus gallus) microbiota following the consumption of zinc biofortified wheat (*Triticum aestivum*)-based diet. J Agric Food Chem 66 (25), 6291–6299.

Reed, S., Neuman, H., Moscovich, S., Glahn, R.P., Koren, O., Tako, E., 2015. Chronic zinc deficiency alters chick gut microbiota composition and function. Nutrients 7 (12), 9768–9784.

Richardson, J.B., Dancy, B.C.R., Horton, C.L., Lee, Y.S., Madejczyk, M.S., Xu, Z.Z., Ackermann, G., Humphrey, G., Palacios, G., Knight, R., Lewis, J.A., 2018. Exposure to toxic metals triggers unique responses from the rat gut microbiota. Sci Rep 8 (1), 6578.

Rodríguez, J.M., Murphy, K., Stanton, C., Ross, R.P., Kober, O.I., Juge, N., Avershina, E., Rudi, K., Narbad, A., Jenmalm, M.C., Marchesi, J.R., Collado, M.C., 2015. The composition of the gut microbiota throughout life, with an emphasis on early life. Microb Ecol Health Dis 26, 26050.

Rosenfeld, C.S., 2015. Microbiome disturbances and autism spectrum disorders. Drug Metab Dispos 43 (10), 1557–1571.

Rothenberg, S.E., Keiser, S., Ajami, N.J., Wong, M.C., Gesell, J., Petrosino, J.F., Johs, A., 2016. The role of gut microbiota in fetal methylmercury exposure: insights from a pilot study. Toxicol Lett 242, 60–67.

Sauer, A.K., Bockmann, J., Steinestel, K., Boeckers, T.M., Grabrucker, A.M., 2019. Altered intestinal morphology and microbiota composition in the autism spectrum disorders associated SHANK3 mouse model. Int J Mol Sci 20 (9).

Sauer, A.K., Grabrucker, A.M., 2019. Zinc deficiency during pregnancy leads to altered microbiome and elevated inflammatory markers in mice. Front Neurosci 13, 1295.

Saurman, V., Margolis, K.G., Luna, R.A., 2020. Autism spectrum disorder as a brain-gut-microbiome Axis disorder. Dig Dis Sci 65 (3), 818–828.

Shao, M., Zhu, Y., 2020. Long-term metal exposure changes gut microbiota of residents surrounding a mining and smelting area. Sci Rep 10 (1), 4453.

Siniscalco, D., Schultz, S., Brigida, A.L., Antonucci, N., 2018. Inflammation and neuro-immune dysregulations in autism spectrum disorders. Pharmaceuticals (Basel) 11 (2).

Smith, J.C., McDaniel, E.G., McBean, L.D., Doft, F.S., Halsted, J.A., 1972. Effect of microorganisms upon zinc metabolism using germfree and conventional rats. J Nutr 102 (6), 711–719.

Tomova, A., Husarova, V., Lakatosova, S., Bakos, J., Vlkova, B., Babinska, K., Ostatnikova, D., 2015. Gastrointestinal microbiota in children with autism in Slovakia. Physiol Behav 138, 179–187.

Vela, G., Stark, P., Socha, M., Sauer, A.K., Hagmeyer, S., Grabrucker, A.M., 2015. Zinc in gut-brain interaction in autism and neurological disorders. Neural Plast 2015, 972791.

Vuong, H.E., Hsiao, E.Y., 2017. Emerging roles for the gut microbiome in autism spectrum disorder. Biol Psychiatry 81 (5), 411–423.

Wei, H., Alberts, I., Li, X., 2013. Brain IL-6 and autism. Neuroscience 252, 320–325.

Warner, B.B., 2019. The contribution of the gut microbiome to neurodevelopment and neuropsychiatric disorders. Pediatr Res 85 (2), 216–224.

Wong, C.P., Rinaldi, N.A., Ho, E., 2015. Zinc deficiency enhanced inflammatory response by increasing immune cell activation and inducing IL6 promoter demethylation. Mol Nutr Food Res 59 (5), 991–999.

Wu, W.L., Hsiao, E.Y., Yan, Z., Mazmanian, S.K., Patterson, P.H., 2017. The placental interleukin-6 signaling controls fetal brain development and behavior. Brain Behav Immun 62, 11–23.

Wu, J., Wen, X.W., Faulk, C., Boehnke, K., Zhang, H., Dolinoy, D.C., Xi, C., 2016. Perinatal lead exposure alters gut microbiota composition and results in sex-specific bodyweight increases in adult mice. Toxicol Sci 151 (2), 324–333.

Yamamoto, M., Khan, N., Muniroh, M., Motomura, E., Yanagisawa, R., Matsuyama, T., Vogel, C.F., 2017. Activation of interleukin-6 and -8 expressions by methylmercury in human U937 macrophages involves RelA and p50. J Appl Toxicol 37 (5), 611–620.

Zhai, Q., Wang, J., Cen, S., Zhao, J., Zhang, H., Tian, F., Chen, W., 2019. Modulation of the gut microbiota by a galactooligosaccharide protects against heavy metal lead accumulation in mice. Food Funct 10 (6), 3768–3781.

CHAPTER 13

Biometal homeostasis as a therapeutic strategy in autism spectrum disorders

Contents

Therapeutic approaches for ASD that target, or influence, metal homeostasis can be quite diverse. For example, influencing the gut microbiome through pre- and probiotics may affect trace metal absorption and bioavailability and, by that, contribute to potentially beneficial effects. Prebiotics are a nondigestible dietary ingredient that selectively stimulates the growth or activity of certain bacteria in the gut, resulting in beneficial effects for the host, while probiotics are specific live microorganisms that populate the gut and thereby alter the microbiota composition after ingestion (Floch, 2014). However, it is currently not known which microbiota may maximize the absorption of essential metals or inhibit the absorption of toxic metals.

Immune modulation and the prevention of maternal immune activation may also prevent the associated changes in metal homeostasis, such as hypozincemia. However, immune activation has an important role in fighting pathogens. Thus, its suppression may be problematic. Therefore, it may be easier to reduce metal imbalances directly by chelating metals that occur in excess and supplementing those that are deficient. This act of balancing metal levels may be performed as a direct response to measured abnormalities or the occurrence of, e.g., maternal infection, or it may be used as a prevention strategy in pregnant women, similar to the supplementation with folic acid that has been introduced to prevent spina bifida.

Biometals in Autism Spectrum Disorders
ISBN 978-0-12-821132-8
https://doi.org/10.1016/B978-0-12-821132-8.00013-6

The use of metal chelation and supplementation in response to altered trace metal levels is dependent on a quick and reliable way to measure the trace metal status of an individual. Unfortunately, this is currently hampered especially by the difficulties in assessing the zinc status during pregnancy. On the other hand, zinc supplementation is not routinely performed during pregnancies. Chelation therapies are, in general, relatively unspecific and are not recommended due to possible side effects. Thus, despite the availability of trace metals as a supplement, the choice of the correct supplement, the mode of delivery, the timepoint of supplementation, and the problems in the assessment of an individual's need for supplementation make influencing biometal homeostasis a difficult task. However, in particular, zinc status seems to be an essential factor concerning ASD.

Controlling zinc status

There are only a few ways to achieve controlled and regional delivery of zinc to specific tissues, and none of them is used regularly in clinical practice. Therefore, trace metal supplementation or chelation occurs on a systemic level, affecting most tissues of the body except for those that are protected by a barrier, such as the brain (blood—brain barrier [BBB]) or, in part, the developing embryo (placenta).

Chelation of zinc is hardly desired, as zinc is an essential metal, and zinc deficiency, not overload, has been linked to ASD in human and animal studies. Therefore, in the majority of cases, the aim is to increase zinc levels by dietary zinc supplementation. This supplementation may be performed in pregnant women as a prevention strategy to avoid prenatal zinc deficiency or in individuals with ASD as a therapeutic strategy to ameliorate the impairments seen in ASD. While the prevention strategy may be more effective, apart from studies using animal models for ASD, no data is available so far from human studies exploring zinc supplementation in pregnant women to reduce the risk for ASD. In contrast, in individuals with ASD, postnatal zinc supplementation showed some effects since many, especially young children with ASD, show low zinc status, which may modify the ASD phenotype. In particular, zinc supplementation may affect a gastrointestinal (GI) pathology that is present as comorbidity in many individuals with ASD.

However, the underlying causes for the low zinc status in both pregnant women and children with ASD may influence the efficacy of zinc supplementation. In pregnant women, more than a low dietary zinc content, the presence of factors inhibiting zinc uptake may increase the risk for zinc deficiency. In particular, the absorption of zinc can be decreased in response to supplements with high iron and/or calcium content and folic acid that are not uncommon to be taken during pregnancy. Besides, a diet rich in phytates may impair zinc uptake (Vela et al., 2015).

In individuals with ASD, zinc absorption may be impaired due to a GI pathology. For example, an altered microbiota composition, and GI surface area (Sauer et al., 2019), as well as abnormal zinc transporter levels may present a significant challenge for zinc supplementation (Pfaender et al., 2017). As a result of both the presence of factors lowering the bioavailability of zinc and the GI pathology that similarly will affect its bioavailability, concentrations of zinc exceeding the Dietary Reference Intake (RDI) may be required. In addition, the type of supplement will influence the bioavailability of zinc, and certain types of supplements may be more suited for use during pregnancy and in individuals with ASD than others (Hagmeyer et al., 2018).

Type of supplement

Zinc supplements can be differentiated into two classes: inorganic zinc supplements (e.g., zinc salts such as zinc oxide, zinc sulfate, zinc acetate, zinc chloride) and organic zinc supplements (e.g., zinc amino-acid complexes). Inorganic zinc salts, once in solution, dissociate quickly, resulting in free zinc ions. These are taken up in the GI tract into enterocytes through zinc importers, in particular, through ZIP4. However, the presence of other metals and folic acid from supplements may interfere with this process, and phytic acid may bind free zinc, thereby making it inaccessible for zinc transporters (Lönnerdal, 2000). This may be especially problematic for pregnant women.

In the case of a diet low in zinc, ZIP4 protein levels increase at the plasma membrane of enterocytes to maximize the absorption of zinc (Dufner-Beattie et al., 2004). However, this mechanism was reported to be affected in some individuals with ASD (Pfaender et al., 2017). Thus, both during pregnancy and in individuals with ASD that have a GI pathology, supplementation with inorganic zinc has limitations. In contrast, organic zinc supplements can be designed to exploit alternative, ZIP4-independent routes of zinc absorption (Sauer et al., 2017). Due to the overall amino acid—like structure, zinc—amino acid complexes (ZnAAs) are taken up by amino acid transporters. At least four sodium-dependent amino-acid transporters and sodium-independent transporters usually mediate the absorption of acidic, basic, and neutral amino acids from the diet (Kandasamy et al., 2018). ZnAAs may be comprised of different amino acids, such as glutamate, methionine, and lysine (ZnGlu, ZnMet, ZnLys), and in combination, they will use different sets of amino acid transporters for both uptake into the enterocyte as well as the passage into the circulation. Since the body has a significantly higher demand for amino acids than for zinc, probably over a thousand-fold, more amino acid transporters are available compared with the number of zinc transporters (Hagmeyer et al., 2018).

In addition of being independent of ZIP4 expression levels, it was also shown that ZnAAs are not affected by common zinc uptake antagonists (calcium/iron, folic acid, phytic acid) in the same magnitude as inorganic supplements (Sauer et al., 2017) and that once in the circulation, ZnAAs complexes are excreted at a slower rate (Maares and Haase, 2020). Although most ZnAAs are not approved for use as a human supplement yet, zinc methionine (ZnMet) is available. Thus, in the future, it will be necessary to carefully test which type of supplement and at what concentration may be most effective for use as possible prevention and treatment strategy of ASD.

Methods of delivery

While supplementation during pregnancy is suitable to ensure adequate zinc status of the mother and the developing offspring, the situation may be different in an individual that already has an ASD. Although dietary zinc supplementation was shown to have some beneficial effects in some studies in the past (see Chapter 7), there is a fundamental flaw in the approach of dietary zinc supplementation, especially for individuals that currently have an adequate zinc status. While zinc supplementation may reduce the effects of zinc deficiency in the GI system or central nervous system (CNS) and thereby modify the ASD phenotype, an individual with adequate zinc status will likely not absorb additional zinc from dietary supplementation and even if, will not transport additional zinc across the BBB. Thus, the effect of zinc supplementation on the ASD phenotype of an individual with adequate zinc status will be limited. Therefore, the zinc status of an individual with ASD needs to be evaluated to understand if zinc delivery into the brain by dietary supplementation can be an effective strategy.

In animal experiments, it was shown that overdoses of zinc elevate serum zinc levels. However, in studies using rodents, even diets with 400%

of the standard zinc content (inorganic zinc) only result in a 20% increase in serum zinc (Grabrucker et al., 2011a). This is likely a consequence of reduced absorption due to the saturation of zinc transporters, such as ZIP4, that mediate zinc uptake in the GI tract (Lönnerdal, 2000). To bypass this effect, some studies performed an intraperitoneal (i.p.) injection of zinc (Baltaci et al., 2005; Sunar et al., 2009; Irato et al., 1996). Nevertheless, the brain is protected by the BBB from excessive zinc that may occur in the circulation. Thus, while studies have shown that a diet low in zinc will eventually lead to a significant reduction in brain zinc levels (Wensink et al., 1987), dietary zinc supplementation hardly leads to an increase in brain zinc concentration.

An increase in brain zinc levels was proposed to be beneficial in ASD (Hagmeyer et al., 2018). If this increase should be realized not during development, but in an individual with functional BBB, other methods need to be developed that can deliver zinc into the CNS. The BBB is comprised of a layer of specialized endothelial cells that interact with different vascular, immune, and neural cells (Daneman and Prat, 2015). Zinc enters the brain at a constant rate if physiological serum zinc concentrations are present. However, outside this range, the rate of zinc transport across the BBB is significantly different (Bobilya et al., 1997). The endothelial cells of the BBB contain distinct types of zinc transporter systems on the apical and basal sides of the cell. These mediate zinc uptake from the serum into the endothelial cell and the subsequent export into the CNS tissue in a process that does not seem to require energy (Bobilya et al., 1997). However, this restrictiveness creates a problem for CNS delivery of zinc using dietary inorganic, and most likely also organic, zinc, especially if the desired final zinc level is above the physiological concentration in the CNS. Thus, the amount of dietary zinc supplementation that is necessary to reach a therapeutic dose in the CNS is hard to predict as individual dietary, environmental, and genetic factors may influence the required dosage. Possibly, the therapeutic concentration may not be reached in the CNS by conventional dietary zinc supplementation.

Unfortunately, to date, no drug exists that can selectively increase brain zinc levels. However, several approaches that may lead to more targeted trace metal interventions have been used. One option is to screen for and utilize compounds that increase uptake of zinc into the brain. For example, already in the 1970s, phenothiazine derivatives such as chlorpromazine, thioridazine, and perphenazine were shown to increase zinc uptake into the brain in zinc-supplemented rodents (Czerniak and Haim, 1971).

Furthermore, n-acetylcysteine may increase glutathione levels (Joshi et al., 2012), which may enhance the delivery of zinc to cells and increase zinc retention in tissues, including the brain. However, this effect is not brain specific.

Apart from altering brain zinc levels indirectly, there are also more direct approaches. A very targeted and brain-specific way to increase zinc is the intrathecal injection of zinc that directly reaches the cerebrospinal fluid (CSF), which freely exchanges zinc with the extracellular fluid of the brain (Larson and Kitto, 1997). In addition, devices such as Alzet pumps implanted in rat brains were able to release up to 1 mM $ZnCl_2$ into the hippocampal hilus for 4 weeks (Elsas et al., 2009). Furthermore, a guide cannula can be implanted into the dorsal hippocampus or lateral ventricle to deliver zinc (Daumas et al., 2004).

Although these techniques are available in rodents, due to their invasiveness, they can hardly be used in humans as a way of zinc delivery to the brain, especially not over a long period.

Recently, novel polymeric nanocarriers that can cross the BBB and encapsulate zinc were developed (Grabrucker et al., 2011a,b). These nanoparticles dramatically improve brain targeting and control over the resulting concentration of zinc in the brain. Once in the brain, the majority is endocytosed and degrades over time, releasing their content into the cell soma. In mice, these nanoparticles were shown to significantly increase brain zinc levels within a few hours after i.p. injection (Chhabra et al., 2015) with no side effects even after chronic injection for 14 days (Vilella et al., 2018).

All these strategies provide opportunities for delivery of zinc into the brain. However, an increase in brain zinc needs to be carefully controlled. Under physiological conditions, free zinc levels are maintained at low levels, and an excess of free zinc is neurotoxic (Weiss et al., 2000; Yokoyama et al., 1986; Choi and Koh, 1998; Choi et al., 1988). Thus, zinc needs to be buffered by proteins such as metallothioneins (MTs) (Aschner, 1996). Approximately 90% of the total brain zinc is bound to proteins (Frederickson, 1989). This binding is reversible, and zinc may be rereleased to act as intracellular signaling ion. For example, zinc is released from MTs in response to nitric oxide (NO) (Lin et al., 2007). Supplemented zinc may trigger an increase in the expression levels of MTs and thereby maximize these intracellular zinc stores (LaRochelle et al., 2008).

Concerning ASD, the role of intracellular zinc at the ASD-associated protein network at the postsynaptic density of synapses may be the most relevant. Thus, zinc must not only be delivered into the brain but also taken up into neurons. Since zinc cannot cross biological membranes by passive diffusion, and zinc transporters might be quickly saturated, zinc delivery by nanocarriers that are endocytosed by cells has another advantage. However, zinc also enters neurons via glutamate receptors (Tóth, 2011) and voltage-gated calcium channels (Atar et al., 1995). So far, little is known about the uptake and role of zinc in glial cells. Nevertheless, the effectiveness of brain zinc delivery in ASD has yet to be proven in preclinical and clinical studies in the future.

Another option to achieve an increase in postsynaptic intracellular zinc is a redistribution of zinc within the cell. Redistribution and chelation of zinc were initially explored in Alzheimer's disease (AD). The use of metal protein attenuating compounds (MPACs) is still among the most promising strategies for the treatment of AD (Sampson et al., 2014; Ritchie et al., 2004). Some of these compounds can scavenge free zinc, for example, from presynaptic vesicles due to their membrane permeability. However, the zinc-loaded compound is then diffusing from the vesicle into the cytosol, where proteins with higher zinc-binding affinity will be able to utilize the zinc ion bound to the compound. For example, the zinc-binding 5-chloro-7-iodo-8-hydroxyquinoline (clioquinol) is a zinc ionophore that can cross the BBB when given at doses of up to 80 mg/day to patients (Regland et al., 2001). In studies using mouse models for ASD, clioquinol improved social interaction (Lee et al., 2015), underlining that altering brain zinc levels, and in particular postsynaptic zinc levels, is a promising strategy for the treatment of ASD in future.

Conclusions

Zinc deficiency during pregnancy should be considered as a significant risk factor for the development of ASD. Since it may be hard to assess the zinc status of pregnant women, general zinc supplementation during pregnancy may be explored as a prevention strategy in the future.

Zinc deficiency is also surprisingly prevalent in individuals with ASD (Grabrucker and Grabrucker, 2017). The precise mechanisms that lead to a decrease in zinc levels are currently not well understood but may be linked to GI disturbances. While a model has been proposed how the effects of zinc deficiency may dysregulate glutamatergic synapses during development and thereby contribute to ASD (Grabrucker, 2014), it is currently not known whether increased zinc levels may be able to reverse the effects later in life. Individuals with ASD that have acute zinc deficiency may benefit in general by improving their health by establishing an adequate zinc status, which may have effects on their ASD behavior. However, it is hard to predict how individuals with ASD who are not zinc-deficient respond to systemic zinc supplementation. In particular, it may be necessary to deliver additional zinc into the brain to reach therapeutic effects, which is prevented by the presence of the BBB.

Some studies that target zinc homeostasis suggest that features of ASD may respond to altered zinc signaling (Lee et al., 2015). However, to maximize the effects, the manipulation of specific brain regions with specific concentrations for a specific amount of time might be necessary. Few technologies to date allow such a precise regulation of local zinc signaling.

Currently, different methods can be employed to alter zinc levels in the body, which may be already useful in the treatment or prevention of ASD, while others need to be fully developed to be applied to humans (Fig. 13.1).

In the future, besides the new development of zinc delivering technologies and zinc supplements, and the identification of compounds that alter zinc levels, it will be necessary to have specific biomarkers for zinc deficiency, especially for mild zinc deficiency that is hard to detect in the analysis of serum. These biomarkers may be used to identify the population of pregnant women that may be at risk of developing zinc deficiency or to identify babies who may benefit from postnatal zinc supplementation. Currently, there are no biomarkers for acute or prenatal zinc deficiency, such as specific microbiota profiles or biochemical, morphological, immunological, and hormonal profiles (Wieringa et al., 2015).

Finally, although zinc supplementation may be useful on its own, a combination of drug treatment together with zinc supplementation to explore synergistic effects has not been investigated in detail in human or animal studies. However, some drugs that are promising in ASD, such as rapamycin (Kotajima-Murakami et al., 2019), 2-methyl-6-phenylethyl-pyridine (MPEP), and memantine (Carlson, 2012) target the same postsynaptic signaling pathways that are also affected by zinc signaling. In addition, the synergistic effects of zinc supplementation and behavioral therapy have not been systematically evaluated so far.

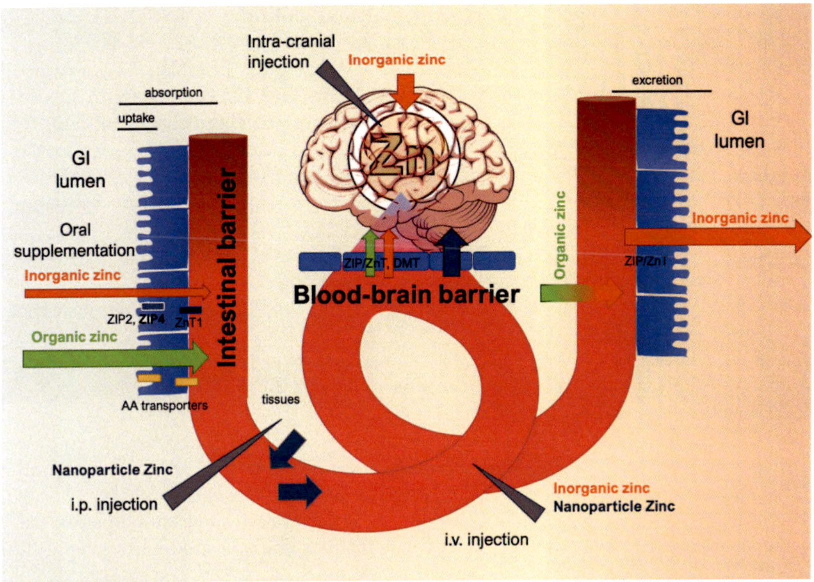

Figure 13.1 Dietary zinc supplementation may be useful to prevent prenatal zinc deficiency and to reduce zinc deficiencies in individuals with ASD. While inorganic zinc supplements are absorbed through zinc transporters (ZIP2, ZIP4 at the apical side of the enterocyte, and ZnT1 at the basal side), organic zinc supplements, i.e., zinc amino acid conjugates are taken up through amino acid transporters. Organic zinc supplements may have an advantage at the intestinal barrier since they are less affected by zinc uptake inhibitors and use less saturable transporters that may be less affected by GI disturbances. However, to reach the brain, inorganic and organic zinc supplements need to cross the BBB, which is a very restricted process. While intrathecal or intracranial delivery of zinc is possible but very invasive, brain targeted zinc delivery by nanocarriers showed promising results in animal studies. These nanoparticles are injected i.p. or i.v. and cross the BBB to release zinc in the brain. Zinc released by nanocarriers or delivered by inorganic and organic supplements will eventually be excreted through feces. However, the retention of organic zinc seems to be higher than that of inorganic zinc supplements. *ASD*, autism spectrum disorder; *BBB*, blood–brain barrier; *GI*, gastrointestinal; *i.p.*, intraperitoneal; *i.v.*, intravenous.

References

Aschner, M., 1996. The functional significance of brain metallothioneins. FASEB J 10 (10), 1129–1136.

Atar, D., Backx, P.H., Appel, M.M., Gao, W.D., Marban, E., 1995. Excitation-transcription coupling mediated by zinc influx through voltage-dependent calcium channels. J Biol Chem 270 (6), 2473–2477.

Baltaci, A.K., Mogulkoc, R., Halifeoglu, I., 2005. Effects of zinc deficiency and supplementation on plasma leptin levels in rats. Biol Trace Elem Res 104 (1), 41–46.

Bobilya, D.J., Guerin, J.L., Rowe, D.J., 1997. Zinc transport across an in vitro blood-brain barrier model. J Trace Elem Exp Med 10, 9–18.

Carlson, G.C., 2012. Glutamate receptor dysfunction and drug targets across models of autism spectrum disorders. Pharmacol Biochem Behav 100 (4), 850–854.

Chhabra, R., Ruozi, B., Vilella, A., Belletti, D., Mangus, K., Pfaender, S., Sarowar, T., Boeckers, T.M., Zoli, M., Forni, F., Vandelli, M.A., Tosi, G., Grabrucker, A.M., 2015. Application of polymeric nanoparticles for CNS targeted zinc delivery in vivo. CNS Neurol Disord Drug Targets 14 (8), 1041–1053.

Choi, D.W., Koh, J.Y., 1998. Zinc and brain injury. Annu Rev Neurosci 21, 347–375.

Choi, D.W., Yokoyama, M., Koh, J., 1988. Zinc neurotoxicity in cortical cell culture. Neuroscience 24 (1), 67–79.

Czerniak, P., Haim, D.B., 1971. Phenothiazine derivatives and brain zinc. Turnover radioactive isotope study. Arch Neurol 24 (6), 555–560.

Daneman, R., Prat, A., 2015. The blood-brain barrier. Cold Spring Harb Perspect Biol 7 (1), a020412.

Daumas, S., Halley, H., Lassalle, J.M., 2004. Disruption of hippocampal CA3 network: effects on episodic-like memory processing in C57BL/6J mice. Eur J Neurosci 20 (2), 597–600.

Dufner-Beattie, J., Kuo, Y.M., Gitschier, J., Andrews, G.K., 2004. The adaptive response to dietary zinc in mice involves the differential cellular localization and zinc regulation of the zinc transporters ZIP4 and ZIP5. J Biol Chem 279 (47), 49082–49090.

Elsas, S.M., Hazany, S., Gregory, W.L., Mody, I., 2009. Hippocampal zinc infusion delays the development of afterdischarges and seizures in a kindling model of epilepsy. Epilepsia 50 (4), 870–879.

Floch, M.H., 2014. Probiotics and prebiotics. Gastroenterol Hepatol (NY) 10 (10), 680–681.

Frederickson, C.J., 1989. Neurobiology of zinc and zinc-containing neurons. Int Rev Neurobiol 31, 145–238.

Grabrucker, A.M., 2014. A role for synaptic zinc in ProSAP/Shank PSD scaffold malformation in autism spectrum disorders. Dev Neurobiol 74 (2), 136–146.

Grabrucker, S., Grabrucker, A.M., 2017. Chapter 9: zinc and autism. In: White, A. (Ed.), Biometals in Neurodegenerative Diseases: Mechanisms and Therapeutics. Academic Press, Cambridge, MA, pp. 153–173.

Grabrucker, A.M., Garner, C.C., Boeckers, T.M., Bondioli, L., Ruozi, B., Forni, F., Vandelli, M.A., Tosi, G., 2011b. Development of novel Zn^{2+} loaded nanoparticles designed for cell-type targeted drug release in CNS neurons: in vitro evidences. PLoS One 6 (3), e17851.

Grabrucker, A.M., Rowan, M., Garner, C.C., 2011a. Brain-delivery of zinc-ions as potential treatment for neurological diseases: mini review. Drug Deliv Lett 1 (1), 13–23.

Hagmeyer, S., Sauer, A.K., Grabrucker, A.M., 2018. Prospects of zinc supplementation in autism spectrum disorders and Shankopathies such as Phelan McDermid syndrome. Front Synaptic Neurosci 10, 11.

Irato, P., Sturniolo, G.C., Giacon, G., Magro, A., D'Inca, R., Mestriner, C., Albergoni, V., 1996. Effect of zinc supplementation on metallothionein, copper, and zinc concentration in various tissues of copper-loaded rats. Biol Trace Elem Res 51 (1), 87–96.

Joshi, D., Mittal, D.K., Shukla, S., Srivastav, A.K., 2012. Therapeutic potential of N-acetyl cysteine with antioxidants (Zn and Se) supplementation against dimethylmercury toxicity in male albino rats. Exp Toxicol Pathol 64 (1–2), 103–108. https://doi.org/10.1016/j.etp.2010.07.001. In this issue.

Kandasamy, P., Gyimesi, G., Kanai, Y., Hediger, M.A., 2018. Amino acid transporters revisited: new views in health and disease. Trends Biochem Sci 43 (10), 752–789.

Kotajima-Murakami, H., Kobayashi, T., Kashii, H., Sato, A., Hagino, Y., Tanaka, M., Nishito, Y., Takamatsu, Y., Uchino, S., Ikeda, K., 2019. Effects of rapamycin on social interaction deficits and gene expression in mice exposed to valproic acid in utero. Mol Brain 12 (1), 3.

LaRochelle, O., Labbé, S., Harrisson, J.F., Simard, C., Tremblay, V., St-Gelais, G., Govindan, M.V., Séguin, C., 2008. Nuclear factor-1 and metal transcription factor-1 synergistically activate the mouse metallothionein-1 gene in response to metal ions. J Biol Chem 283 (13), 8190–8201.

Larson, A.A., Kitto, K.F., 1997. Manipulations of zinc in the spinal cord, by intrathecal injection of zinc chloride, disodium-calcium-EDTA, or dipicolinic acid, alter nociceptive activity in mice. J Pharmacol Exp Ther 282 (3), 1319–1325.

Lee, E.J., Lee, H., Huang, T.N., Chung, C., Shin, W., Kim, K., Koh, J.Y., Hsueh, Y.P., Kim, E., 2015. Trans-synaptic zinc mobilization improves social interaction in two mouse models of autism through NMDAR activation. Nat Commun 6, 7168.

Lin, W., Mohandas, B., Fontaine, C.P., Colvin, R.A., 2007. Release of intracellular Zn(2+) in cultured neurons after brief exposure to low concentrations of exogenous nitric oxide. Biometals 20 (6), 891–901.

Lönnerdal, B., 2000. Dietary factors influencing zinc absorption. J Nutr 130 (5S Suppl.), 1378–1383.

Maares, M., Haase, H., 2020. A guide to human zinc absorption: general overview and recent advances of in vitro intestinal models. Nutrients 12 (3).

Pfaender, S., Sauer, A.K., Hagmeyer, S., Mangus, K., Linta, L., Liebau, S., Bockmann, J., Huguet, G., Bourgeron, T., Boeckers, T.M., Grabrucker, A.M., 2017. Zinc deficiency and low enterocyte zinc transporter expression in human patients with autism related mutations in SHANK3. Sci Rep 7, 45190.

Regland, B., Lehmann, W., Abedini, I., Blennow, K., Jonsson, M., Karlsson, I., Sjögren, M., Wallin, A., Xilinas, M., Gottfries, C.G., 2001. Treatment of Alzheimer's disease with clioquinol. Dement Geriatr Cogn Disord 12 (6), 408–414.

Ritchie, C.W., Bush, A.I., Masters, C.L., 2004. Metal-protein attenuating compounds and Alzheimer's disease. Expert Opin Investig Drugs 13 (12), 1585–1592.

Sampson, E.L., Jenagaratnam, L., McShane, R., 2014. Metal protein attenuating compounds for the treatment of Alzheimer's dementia. Cochrane Database Syst Rev (2), CD005380.

Sauer, A.K., Bockmann, J., Steinestel, K., Boeckers, T.M., Grabrucker, A.M., 2019. Altered intestinal morphology and microbiota composition in the autism spectrum disorders associated SHANK3 mouse model. Int J Mol Sci 20 (9).

Sauer, A.K., Pfaender, S., Hagmeyer, S., Tarana, L., Mattes, A.K., Briel, F., Küry, S., Boeckers, T.M., Grabrucker, A.M., 2017. Characterization of zinc amino acid complexes for zinc delivery in vitro using Caco-2 cells and enterocytes from hiPSC. Biometals 30 (5), 643–661.

Sunar, F., Baltaci, A.K., Ergene, N., Mogulkoc, R., 2009. Zinc deficiency and supplementation in ovariectomized rats: their effect on serum estrogen and progesterone levels and their relation to calcium and phosphorus. Pak J Pharm Sci 22 (2), 150–154.

Tóth, K., 2011. Zinc in neurotransmission. Annu Rev Nutr 31, 139–153.

Vela, G., Stark, P., Socha, M., Sauer, A.K., Hagmeyer, S., Grabrucker, A.M., 2015. Zinc in gut-brain interaction in autism and neurological disorders. Neural Plast 2015, 972791.

Vilella, A., Belletti, D., Sauer, A.K., Hagmeyer, S., Sarowar, T., Masoni, M., Stasiak, N., Mulvihill, J.J.E., Ruozi, B., Forni, F., Vandelli, M.A., Tosi, G., Zoli, M., Grabrucker, A.M., 2018. Reduced plaque size and inflammation in the APP23 mouse model for Alzheimer's disease after chronic application of polymeric nanoparticles for CNS targeted zinc delivery. J Trace Elem Med Biol 49, 210–221.

Weiss, J.H., Sensi, S.L., Koh, J.Y., 2000. Zn(2+): a novel ionic mediator of neural injury in brain disease. Trends Pharmacol Sci 21 (10), 395–401.

Wensink, J., Lenglet, W.J., Vis, R.D., Van den Hamer, C.J., 1987. The effect of dietary zinc deficiency on the mossy fiber zinc content of the rat hippocampus. A microbeam PIXE study. Particle induced X-ray emission. Histochemistry 87 (1), 65–69.

Wieringa, F.T., Dijkhuizen, M.A., Fiorentino, M., Laillou, A., Berger, J., 2015. Determination of zinc status in humans: which indicator should we use? Nutrients 7 (5), 3252–3263.

Yokoyama, M., Koh, J., Choi, D.W., 1986. Brief exposure to zinc is toxic to cortical neurons. Neurosci Lett 71 (3), 351–355.

CHAPTER 14

Future perspectives: autism, a disorder of biometal imbalance?

Contents

ASDs are neurodevelopmental disorders with a strong genetic component. Especially, genome-wide association studies with large cohorts of individuals with ASDs have revealed a high number of genes that are linked to ASDs as causative factors (Bourgeron, 2016). By far, not every function of each ASD-associated gene has been described. However, it was noted that pathogenic mutations frequently occur in genes coding for proteins that are found within a shared signaling network at synapses in the central nervous system (CNS) (Kelleher et al., 2012; Huguet et al., 2016). Nevertheless, despite the identification of this synaptic pathway, there is considerable heterogeneity of identified candidate genes. Many of them seem not to be linked to synaptic processes, which imposes a significant challenge to define a common theme.

ASDs are neurodevelopmental disorders that also have a strong environmental component. The nongenetic risk factors for ASDs may act independent of genetic factors, but likely, there is a complex interplay of genetic and nongenetic risk factors in the etiology of ASDs. Among the major nongenetic risk factors are maternal immune activation, maternal zinc deficiency, and in utero exposure to certain drugs/toxins. These seemingly unrelated factors further compound the identification of a common theme in ASDs and to propose a unified theory that encompasses genetic as well as environmental factors in a mechanistic explanation of ASDs.

Biometals in Autism Spectrum Disorders
ISBN 978-0-12-821132-8
https://doi.org/10.1016/B978-0-12-821132-8.00014-8

Abnormal trace metal biology, which includes the dysfunction of metal-binding proteins and their associated biological processes, imbalances in metal homeostasis, and the effects on proteins, and abnormal metal ion signaling, may indeed be an excellent candidate to unify several risk factors of ASDs in one common theme.

Trace metal imbalances in autism spectrum disorder

Chapter 3 summarizes data that show that abnormal trace metal levels are common in individuals with ASD. The most reliable link is found between ASD and overload with lead and high levels of copper, together with zinc deficiency. Given that trace metals interact, it may not be possible to distinguish if this ASD-associated trace metal profile is caused by zinc deficiency, which leads to a subsequent increase in lead and copper levels, or by copper or lead overload that reduces zinc levels. However, the relatively high incidence rate of zinc deficiency (Wessells and Brown, 2012), and especially mild subclinical transient zinc deficiency, makes it likely that in the majority of cases, a low bioavailability of zinc, whether caused through dietary factors, gastrointestinal disturbances, or increased demand during pregnancy, is the driving force in the establishment of this profile.

As discussed in Chapters 4 and 5, this ASD-associated trace metal profile will increase oxidative stress in the developing brain and facilitate proinflammatory signaling. In Chapters 8 and 9, data have been highlighted that show that in mouse models, prenatal zinc deficiency indeed induces ASD-like behavior, and a mechanism has been proposed how a lack of zinc may affect synapses through dysregulation of the ASD-associated SHANK2 and SHANK3 proteins.

Trace metals and altered gut–brain signaling in autism spectrum disorder

In Chapter 12, data are presented that show that an imbalance in trace metals such as zinc deficiency or overload with lead can induce alterations in the gastrointestinal (GI) tract. In particular, an altered microbiota composition that has been reported in many human studies and animal models for ASD is induced (Nitschke et al., 2020). In addition, some ASD-associated proteins are found in the GI system and also cause a GI pathology that may be linked to impaired zinc uptake (see Chapter 11). Together, these GI disturbances may lead to an impaired intestinal permeability (leaky gut), which activates the immune system and leads to the release of inflammatory cytokines, which in return, also affect zinc homeostasis. Thus, the ASD-associated trace metal profile is connected to both a pathology in the CNS and a GI pathology, which mirrors the situation in humans with ASD, who frequently have GI problems (Kostiukow et al., 2020).

Trace metals and immune activation in autism spectrum disorder

The immune activation that may occur through the GI pathology with alterations of the gut microbiota composition and intestinal permeability can also be modeled directly through inducing an immune response in pregnant mice. This maternal immune activation (MIA) is a model for prenatal infections that have been linked to ASD in human studies. In Chapter 10, data are presented that show a link between MIA and hypozincemia. Also MIA alters gut microbiota composition (Li et al., 2018). Both GI pathologies and MIA through injection of LPS or Poly I:C leads to an upregulation of proinflammatory cytokines that has been observed in many individuals with ASD (Gottfried et al., 2015).

Thus, the ASD-associated trace metal profile can induce a GI pathology and immune activation, but a GI pathology and immune activation will further decrease zinc levels, resulting in a vicious cycle where these factors in ASD are linked in a common theme. If the cycle is broken by zinc supplementation, as shown in rats with MIA, ASD-like behavior is abolished (Kirsten et al., 2015).

Trace metals and other nongenetic factors in autism spectrum disorder

Following this model, other nongenetic factors and genetic factors should converge on the induced GI or CNS pathology, or both. Another prominent nongenetic model for ASD is the induction of ASD-like behavior in animals through in utero exposure to valproic acid (VPA). VPA-exposed models indeed also show GI disturbance and an altered microbiota composition (Liu et al., 2018) and abnormal zinc homeostasis. Again, zinc supplementation can prevent ASD-like behaviors in VPA-exposed rats (Cezar et al., 2018).

Trace metals and genetic factors in autism spectrum disorders

ASD-associated genes encode for a variety of proteins. Many of these proteins are linked to synaptic signaling processes and can be found within the same protein complex centered around SHANK proteins (Guang et al., 2018). Chapter 8 explains how the ASD-associated trace metal profile and zinc deficiency, in particular, can modulate the ASD-associated signaling network at synapses and thereby induces a synaptopathy. One could argue that indeed, the proteins are part of a zinc signaling pathway and that equally, the lack of zinc and mutations in proteins in this pathway causes a disruption that is linked to ASD. The close link between the synaptic pathway and zinc can be seen through the possibility to prevent ASD-like behavior in *Shank3* knockout mice through zinc supplementation (Fourie et al., 2018).

Similarly, ASD-associated proteins outside this synaptic pathway that so far could not easily be integrated into a common theme in ASD may be part of trace metal—regulated processes or regulate trace metal homeostasis, and their mutation may affect one or more aspects of the many pathways in the body that are controlled by trace metals. For example, other ASD candidate genes can be indeed linked to the ASD-associated trace metal profile by acting themselves as proteins regulating trace metal levels (discussed in Chapter 8). In the future, all ASD-associated genes should be investigated whether they are zinc-binding or regulated by zinc-dependent processes.

The many different risk factors for ASD are reflected by the various animal models for ASD that have been generated in the past (Ueoka et al., 2019; Verma et al., 2019). Among the genetic mouse models for ASD are *Integrin β3, Neuroligin (Nlgn) 1, Nlgn2, Nlgn3, Nlgn4, Neurexin1α, Contactin-associated protein2 (Cntnap2), Shank2, Shank3, Engrailed-2 (En2), Forkhead box protein 2 (Foxp2), Phosphatase and tensin homolog (Pten), Glutamate [NMDA] receptor subunit epsilon-2 (Grin2B), Synaptic Ras GTPase Activating Protein 1 (SynGAP1)*, and *Ubiquitin-protein ligase E3A (Ube3A)* transgenic mice (Crawley, 2012). These mice have mutations in genes that are all linked to the zinc-dependent synaptic signaling pathway mediated by SHANK2 and SHANK3 signaling and dynamics. *Chromodomain helicase DNA-binding protein 8 (Chd8)* encodes for a zinc-binding protein. Transgenic mice have multiple deficits due to the more general role of CHD8 in histone and chromatin modification (Platt et al., 2017) that also include

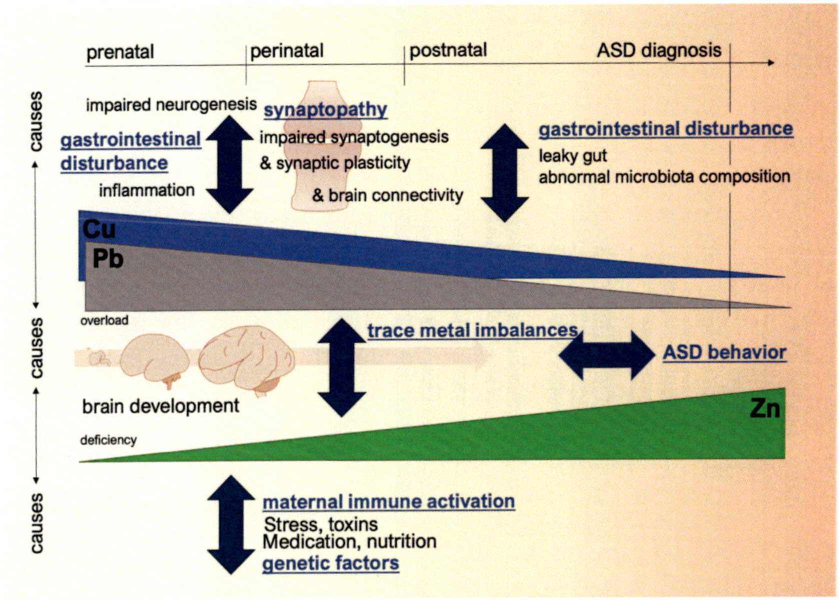

Figure 14.1 Trace metal imbalances unify several causes for ASD. An increase in copper (Cu) and (Pb), and the concurrent decrease in zinc (Zn) (and vice versa) is linked to maternal immune activation. On the one hand, maternal immune activation causes trace metal imbalances; on the other hand, trace metal imbalances cause gastrointestinal disturbances such as a leaky gut and altered microbiota composition that may lead to immune activation. Together, immune activation and gastrointestinal problems further aggravate metal dysregulation. The resulting increase in inflammatory cytokines and zinc deficiency affects brain development by impairing neurogenesis, synaptogenesis, and synapse function, which ultimately results in abnormal brain connectivity. At the time of the ASD diagnosis, trace metal imbalances and maternal immune activation may no longer be detected, but the altered brain connectivity may have lasting consequences visible in the ASD-associated behaviors. Several other risk factors for ASD can be integrated into this model, such as prenatal stress and exposure to toxins that lead to immune activation, or medications such as valproic acid that affect trace metal homeostasis. Genetic factors such as ASD-associated genes that are found at the synapse are regulated by zinc-dependent SHANK2 and SHANK3 proteins and may be involved in other zinc-dependent processes. Therefore, their mutations may mimic zinc deficiency. Other ASD candidate genes encode for proteins that mediate metal homeostasis directly.

altered synaptic physiology. Among the nongenetic mouse models are MIA, VPA, and prenatal zinc-deficient mice. Their pathology is also linked to abnormal trace metal status. While the ASD-related behaviors of MIA, VPA, *Shank2,* and *Shank3* KO mice can be modified by zinc supplementation, less is known about the effects of zinc in other model systems.

Conclusions

The data described above show how the ASD-associated trace metal profile can induce key features described in humans with ASD and ASD animal models that, in their completeness, cannot be modeled through other risk factors of ASD. Trace metal imbalances can explain the presence of increased oxidative stress, a synaptopathy, GI disturbances, altered microbiota profiles, and increased inflammation. Thus, MIA, the use of certain medications, zinc deficiency, overload with toxic metals, and major ASD-associated genes whose dysfunction may mimic impaired zinc signaling, are linked in a single pathomechanism: **trace metal imbalance** (Fig. 14.1).

The findings to date highlight the importance of the science of biometals in ASD. In the future, more research on ASD will hopefully consider this vital aspect of the disorders, which may lead to new mechanistic insights and biomarkers, as well as preventive and therapeutic strategies.

References

Bourgeron, T., 2016. Current knowledge on the genetics of autism and propositions for future research. C R Biol 339 (7–8), 300–307.

Cezar, L.C., Kirsten, T.B., da Fonseca, C.C.N., de Lima, A.P.N., Bernardi, M.M., Felicio, L.F., 2018. Zinc as a therapy in a rat model of autism prenatally induced by valproic acid. Prog Neuro-Psychopharmacol Biol Psychiatry 84 (Pt A), 173–180.

Crawley, J.N., 2012. Translational animal models of autism and neurodevelopmental disorders. Dialogues Clin Neurosci 14 (3), 293–305.

Fourie, C., Vyas, Y., Lee, K., Jung, Y., Garner, C.C., Montgomery, J.M., 2018. Dietary zinc supplementation prevents autism related behaviors and striatal synaptic dysfunction in Shank3Exon 13-16 mutant mice. Front Cell Neurosci 12, 374.

Gottfried, C., Bambini-Junior, V., Francis, F., Riesgo, R., Savino, W., 2015. The impact of neuroimmune alterations in autism spectrum disorder. Front Psychiatr 6, 121.

Guang, S., Pang, N., Deng, X., Yang, L., He, F., Wu, L., Chen, C., Yin, F., Peng, J., 2018. Synaptopathology involved in autism spectrum disorder. Front Cell Neurosci 12, 470.

Huguet, G., Benabou, M., Bourgeron, T., 2016. The genetics of autism spectrum disorders. In: Sassone-Corsi, P., Christen, Y. (Eds.), A Time for Metabolism and Hormones. Springer, Cham (CH).

Kelleher, R.J., Geigenmüller, U., Hovhannisyan, H., Trautman, E., Pinard, R., Rathmell, B., Carpenter, R., Margulies, D., 2012. High-throughput sequencing of mGluR signaling pathway genes reveals enrichment of rare variants in autism. PLoS One 7 (4), e35003.

Kirsten, T.B., Queiroz-Hazarbassanov, N., Bernardi, M.M., Felicio, L.F., 2015. Prenatal zinc prevents communication impairments and BDNF disturbance in a rat model of autism induced by prenatal lipopolysaccharide exposure. Life Sci 130, 12–17.

Kostiukow, A., Poniewierski, P., Daroszewski, P., Samborski, W., 2020. Gastrointestinal disorders in children with autism spectrum disorder. Pol Merkur Lek 48 (283), 69–72.

Li, Y., Dugyala, S.R., Ptacek, T.S., Gilmore, J.H., Frohlich, F., 2018. Maternal immune activation alters adult behavior, gut microbiome and juvenile brain oscillations in ferrets. eNeuro 5 (5).

Liu, F., Horton-Sparks, K., Hull, V., Li, R.W., Martínez-Cerdeño, V., 2018. The valproic acid rat model of autism presents with gut bacterial dysbiosis similar to that in human autism. Mol Autism 9, 61.

Nitschke, A., Deonandan, R., Konkle, A.T., 2020. The link between autism spectrum disorder and gut microbiota: a scoping review. Autism, 1362361320913364.

Platt, R.J., Zhou, Y., Slaymaker, I.M., Shetty, A.S., Weisbach, N.R., Kim, J.A., Sharma, J., Desai, M., Sood, S., Kempton, H.R., Crabtree, G.R., Feng, G., Zhang, F., 2017. Chd8 mutation leads to autistic-like behaviors and impaired striatal circuits. Cell Rep 19 (2), 335–350.

Ueoka, I., Pham, H.T.N., Matsumoto, K., Yamaguchi, M., 2019. Autism spectrum disorder-related syndromes: modeling with Drosophila and rodents. Int J Mol Sci 20 (17).

Verma, V., Paul, A., Amrapali Vishwanath, A., Vaidya, B., Clement, J.P., 2019. Understanding intellectual disability and autism spectrum disorders from common mouse models: synapses to behaviour. Open Biol 9 (6), 180265.

Wessells, K.R., Brown, K.H., 2012. Estimating the global prevalence of zinc deficiency: results based on zinc availability in national food supplies and the prevalence of stunting. PLoS One 7 (11), e50568.

Further reading

Nakanishi, M., Anderson, M.P., Takumi, T., 2019. Recent genetic and functional insights in autism spectrum disorder. Curr Opin Neurol 32 (4), 627–634.

Index

'*Note:* Page numbers followed by "f" indicate figures and "t" indicates tables.'

H

I

L

M

Kolibri 360 GmbH
Ferdinand-Jühlke-Straße 7
99095 Erfurt, Deutschland
produktsicherheit@kolibri360.de